Ponderosa
602 268 1261

S. Mt
602 276 8131

Beyond Sun and Sand

Beyond Sun and Sand

CARIBBEAN ENVIRONMENTALISMS

EDITED BY

SHERRIE L. BAVER

BARBARA DEUTSCH LYNCH

Rutgers University Press
New Brunswick, New Jersey, and London

Library of Congress Cataloging-in-Publication Data

Beyond sun and sand : Caribbean environmentalisms / edited by Sherrie L. Baver and
Barbara Deutsch Lynch.
    p. cm.
    Includes bibliographical references and index.
    ISBN-13: 978-0-8135-3653-8 (hardcover : alk. paper)
    ISBN-13: 978-0-8135-3654-5 (pbk : alk. paper)
  1. Environmental policy—Caribbean Area. 2. Environmentalism—Caribbean Area.   I.
Baver, Sherrie L. II. Lynch, Barbara D.
  GE190.C27B49   2005
  333.72'09729—dc21

2005004406

A British Cataloging-in-Publication record for this book is available from the British Library

Manufactured in the United States of America

CONTENTS

ACKNOWLEDGMENTS

As usual, numerous people helped in the long germination and publication of this book; we can mention only a few of them by name. Sherrie would like to thank several supportive colleagues in New York, Puerto Rico, and the Dominican Republic, especially Janis Roze, who introduced her to environmental issues over a decade ago, Manny Ness and Pedro Caban. Her family, Chris and Nicholas, continue to be supportive and patient. Barbara would like to thank her husband, Paul, her kids, the Río Yaque del Norte and the memory of Redbud Woods. She also would like to recognize two people in particular, who helped shape her thinking, Pedro Juan del Rosario and Maria Caridad Cruz. Finally, both editors want to thank the contributors to the volume, and in the case of the late Marion Miller, her family, for their help with the details involved in putting together an edited volume. We are also grateful to Andy Love at Cornell and the staff at Rutgers University Press, especially Kristi Long, Nicole Manganaro, and Evan Young for all of their efforts.

PART I

Issues and Movements

CHAPTER 1

The Political Ecology of Paradise

SHERRIE L. BAVER AND BARBARA DEUTSCH LYNCH

Introduction

Filtered through the lens of the European and North American media, the Caribbean becomes a series of uniformly breezy landscapes of sun and sand designed for loafing, sailing, diving, and perhaps for gambling and sex. In the conservation literature, Caribbean landscapes are habitat for endangered coral reefs and their denizens, parrots, butterflies, caiman, snails and whales and myriad plant species. In either version, the idyllic island landscape is a screen that conceals worlds that are far richer culturally, but trapped in a global economy that offers few options for development. The islands of the Greater and Lesser Antilles are linguistically and culturally diverse and their governments differ in form and in capacity, but they share a history of colonization, demographic transformation through labor migration, and economic dependency on activities that have utterly transformed their landscapes—plantation agriculture, mining, and tourism. It is in this context that environmental policy makers and activists are responding to existing threats and seeking to protect a natural patrimony that is also an economic lifeline.

The landscape as screen also conceals the development policy choices Caribbean governments are making, the environmental consequences of those choices, and the transnational connections that are reshaping the islands. At the same time that multinational corporations are moving operations to islands where financial laws are lax and wages low, large numbers of Caribbean people have moved away to the north where many continue to make vital contributions—economic, social, cultural, and political—to their places or origin. They have done much to shape the North American environmental justice movement by bringing Caribbean environmental sensibilities, political institutions, and organizational traditions to environmental struggles in the United States and Canada. In this volume we examine the environmental dilemmas that face those who live in these worlds behind the screen, the tough environmental choices that must be made within that world, and the strategies and tactics employed by activists who would reconcile the idyllic and the real.[1]

The idea for this volume grew out of a conference on "Environmental Issues in the Caribbean and Caribbean Diaspora" held at the City University of New York. The conference, supported by the Ford Foundation, brought together scholars, development practitioners, NGO representatives, and community activists to begin a conversation on Caribbean environment that would break down linguistic and cultural barriers that have divided the English, French, and Spanish-speaking islands from one another. The conference also sought to bring together scholars and activists, and several essays were written by Caribbean scholars who have played active roles in national and regional environmental debates. We also felt it important to represent the diaspora at the conference and in this volume. Chapters on immigrant communities in New York City highlight similarities in the environmental challenges faced in both worlds and in the strategies used to overcome them.

International Institutions and the Caribbean Environmental Agenda

Once viewed as a luxury of the industrialized world, environmental concern grew in the island Caribbean in the 1980s and emerged as a distinct development policy arena following publication of the Brundtland Commission Report (WCED1988) and the 1992 United Nations Conference on Environment and Development that produced "Agenda 21." These two U.N. initiatives, which sought to reconcile the often conflicting imperatives of environmental protection and economic growth, played an important role in shaping Caribbean environmental agendas.

More often than not, Caribbean environmental policy makers have bought into the elusive concept of "sustainable human development," a concept made popular in international development circles in the 1980s and 1990s (Navajas et al. 1997, ch. 1). The term "sustainability" is problematic: it addresses the intergenerational concerns of environmentalists, but leaves open the questions of what is to be sustained and for whose benefit. As used by the United Nations Development Program (UNDP), the term conveys an approach that privileges natural resource management, biodiversity conservation, and pollution reduction over other environmental issues, even though it emphasizes the importance of such social measures such as reducing poverty, enhancing women's participation, and strengthening institutions.

The Brundtland Report (WCED 1987) and Agenda 21, like other United Nations documents, were products of considerable North-South dialogue. Still, it has been argued that the definition of environmental protection that underlies them was imposed on the Caribbean region by donors from the north. In the intervening years, southern environmental agendas have become highly sophisticated and increasingly focused on the need to tie change in consumption patterns in the north to environmental sustainability in the rest of the world and on the need to address directly the environmental challenges faced by poor urban dwellers. This logic is reflected in the Millennium Development Goals (UNDP 2003) and in the Caribbean Environmental Outlook (Walling et al. 2004). However, the Millennium Development Goals do little to address the contributions of the global north to environmental degradation. Rather, they focus on poverty in the global south as the primary cause of environmental harm and set out development objectives that differ little from those of earlier UN and World Bank statements. The Caribbean Environmental Outlook attributes the region's environmental problems to social and economic vulnerabilities that are at least in part caused by its weak position in the global economy, yet it too defines its policy agenda as if the global economy were an insignificant contributor to environmental degradation. While all of these documents specify that environmental protection is as important for the nations of the south as it is for the nations of the north, they have provoked new debates about whose priorities should determine Caribbean environmental agendas and about who should pay for protection and remediation.

A 1997 UNDP assessment of environmental problems in Latin America and the Caribbean describes the region as "a paradox of abundance and poverty" (Navajas et al. 1997, 7). According to the report, the Caribbean shares with the rest of the region a common set of problems: deforestation (to the point of desertification in parts of Haiti); transition of agriculture to ranching or input-intensive export-oriented agriculture; rapid urbanization coupled with air and water pollution; and the destruction of coastal ecosystems. Because of their small size and relative poverty, the Caribbean islands are particularly vulnerable to environmental degradation.

Like the Brundtland Commission findings, the UNDP assessment acknowledges the link between poverty and environmental degradation. Caribbean citizens and environmental activists are particularly concerned about localized yet extreme problems of water and air pollution and solid waste mishandling—highly visible manifestations of the absence of adequate legal and regulatory frameworks and enforcement mechanisms. To use Beck's (1992) terminology, Caribbean societies have become "risk societies." Indeed, some of the most glaring problems facing the Caribbean in the 1990s—unsafe drinking water, indiscriminate use of pesticides, unregulated industrial emissions, and unrestricted shoreline development—are issues that have been addressed at least to some degree in North America and Western Europe over several decades of judicial and regulatory activity.

Whether the strategies employed in the north are feasible for poor nations whose options are limited in the current global economic context remains an open question. In the United States, for example, first-generation environmentalism was generally pollution-driven, and problems were defined by environmental media—air, water, and waste. Redress was sought through a complex structure of separate (and sometimes conflicting) laws and detailed regulations (Esty and Chertow 1997, Introduction). Regulatory approaches to environmental management are regularly challenged not only in U.S. policy arenas, but also within the context of regional and international trade agreements. Second-generation approaches to environmental policy making have emphasized cooperation among stakeholders, and comprehensive rather than segmented policies. The environmental standards adopted by the International Organization for Standards (ISO) is a prime example of this kind of incentive-based, non-confrontational approach to environmental protection. Preliminary efforts at self-regulation by industries in Mexico (Lezama 2000), the Dominican Republic (Lynch 2001), and Brazil (Roberts and Thanos 2003) have not been encouraging. Where does this leave the Caribbean? Regulation, monitoring, and enforcement are important tools for environmental protection, yet small island nations—and worse yet, municipal authorities within them—rarely enjoy the resources these activities require. All too often, where regulations and structures for enforcement have been put into place, they serve the interests of resource, tourism, and real estate developers at the expense of ordinary citizens. Similarly, even where they are included in stakeholder dialogues, residents of poor urban settlements have little input into problem definition. In this context, NGO and social movement activity can play a significant role in promoting environmental justice agendas.

The chapters in this volume generally approach environmental questions from a political ecology perspective. According to Peet and Watts (2004, 4), political ecology examines "the complex relation of Nature and Society through careful analysis of social forms of access and control over resources—with all their implications for environmental health and sustainable livelihoods." The political ecology approach emphasizes the influence of global capitalism on the production of landscapes as well as the rise of social movements that link economic and environmental justice concerns with concerns for human rights and cultural identity. Some scholars working within the political ecology paradigm ask questions about the distribution of the costs not only of resource depletion and pollution, but of environmental protection and mitigation within and among nations.[2] Others, like Goldman (2001a), ask about how ecological problems are defined in international development contexts and about the kinds of policies that flow from these definitions.

We see a political ecology perspective as a useful entry point from which to understand the ecological problems of the Caribbean and the Caribbean diaspora. It allows us to look at the evolution of Caribbean environments in a regional historical context and at the distribution of environmental ills as a function of global and national political economies. The political ecology approach also allows us to see how policy actors in different contexts have responded to these problems and with what consequences for different social groups and regions.

political ecology

Actors in Caribbean Environmental Struggles

One way to approach this collection is to examine the actors engaged in environmental policy making. The countries of the global south, including the Caribbean, are at a disadvantage vis-à-vis the north in environmental protection as well as other policy areas. However, the islands' differing histories and political systems have resulted in very different national policymaking structures. Furthermore, different sets of actors have had varying degrees of influence on environmental policy depending upon the country or the issue area. The essays in this volume reveal the interactions among three key sets of actors: states, the private sector, and loose alliances of transnational and national nongovernmental organizations (NGOs).

States

While these essays do not systematically examine states' roles in environmental policy making, they all address the ways national development policies have contributed to the present environmental predicament. Caribbean states enjoy differing degrees of autonomy. In Cuba, for example, the state has the overwhelming preponderance of authority in environmental decision making (Diíaz-Briquets and Pérez-López 2000). The Dominican political system has been characterized as "neopatrimonial" (Hartlyn 1998). Environmental policymaking authority rests with the executive, reflecting a broader tendency toward presidentialist rule. In the two territories discussed in this volume, Puerto Rico and Martinique, we find a greater role for legislatures and courts in the shaping of environmental policy and a less dominant executive. While it is probably true that citizens of Martinique and Puerto Rico have more channels through which to pursue environmental grievances than their counterparts in sovereign Caribbean states, it is not at all clear that they have as much leverage as their counterparts in France and the United States.

Although weakened considerably by the structural adjustment programs imposed by international lending institutions, Caribbean states remain key actors in environment and development policy making—in part because international institutions see the Third World states as playing a key role in natural resource management. For example, the 1997 *World Development Report*, issued by the World Bank (1997, 4), cites environmental protection as one of five fundamental tasks that "lie at the core of every government's mission, without which sustainable, shared, poverty-reducing development is impossible." This is especially true for states whose revenues are largely dependent on tourism.

The Private Sector

Dating back to the Caribbean Basin Initiative and the structural adjustment programs of the 1980s, development policies in the island Caribbean (including Cuba)—influenced by the neoliberal paradigm—have accorded heavy weight to interests of multinational corporations. Transnational capital has been a key element in the development of three sectors, each with pronounced environmental impacts: plantation agriculture, export-platform industrialization, and tourism.

Because it implies a controlled approach that is sensitive to local surroundings and communities, many environmentalists regard the tourism industry as the best hope for attracting foreign and domestic private-sector actors capable of responding to an ethos of environmental stewardship as well as to market-based incentives. Much of the debate over the shape and power of the tourist industry plays out in the public/private property controversies that are fundamental to the environmental politics in the region. Idyllic landscapes are created and maintained by excluding local residents from large portions of the islands.

Access to coastal zones has become a key environmental issue as developers have sequestered Caribbean scenic areas for tourist facilities and kept local residents away from beaches they once used for fishing and recreation. Assignment of protected area status to large portions of the islands—encouraged by international institutions—is also controversial. When parks and reserves are established to reinforce the edenic illusion for tourism, attention is seldom paid to the consequences for local residents whose livelihoods depend upon the sequestered landscape. In Sheller's words, "the picturesque vision of the Caribbean continues to be a form of world-making which allows tourists to move through the Caribbean and see Caribbean people simply as scenery" (2003, 63). Throughout the region, governments are promoting ecotourism as an alternative to mass tourism. It is nowhere clear, however, what ecotourism means, what percentage of the tourism industry it can be reasonably expected to encompass, and what difference it will make for local economies and societies.[3]

The material and moral incentives for environmental stewardship in other private-sector activities common to the region—mining, agriculture, cattle raising, and export platform manufacturing—have been historically weak. However, some multinational corporations and producers for consumption in Europe and the United States have felt pressure to behave in a more environmentally responsible fashion coming from consumers or from the need to contain costs. Surprisingly, since the 1990s, the most rapid environmental progress has taken place in the agricultural sector, where plummeting sugar and coffee prices, white fly infestations, and foreign exchange scarcities have spurred a shift away from conventional agriculture toward more sustainable alternatives. While Cuba played a leading role in the adoption of organic practices, other countries have followed suit. The Dominican Republic, for example, has become the region's largest producer of organic bananas; one of its major ice cream manufacturers has entered into agreements with small cacao and coffee producers to supply organic ingredients for its premium products (see Chapter 7 in this volume).

Manufacturers' responses to environmental pressures are less promising. The mining industry continues to be a significant source of pollution as well as landscape transformation, and mining industry environmental efforts may be little more than "greenwashing." One Canadian mining company operating in the Dominican Republic created a foundation that has supported small reforestation and community environmental projects that are unrelated to its own activities (Falconbridge 2005); a second hired consultants to carry out impact assessments, but it was not clear how the results of assessments would be used. Meanwhile growing piles of mine tailings continue to contaminate rivers, rendering agricultural lands uncultivable. In Cuba and the Dominican Republic, some of the worst polluting manufacturing plants of the import substitution era have been shut down, and others have been required to install treatment equipment, but pollution abatement efforts in export platform industrial zones throughout the region have been weak, and are complicated by the fact that park owners rather than manufacturers are held responsible for sewage treatment. Few if any Caribbean islands have sewage and solid waste treatment facilities adequate for handling industrial waste.

Civil Society: NGOs, Community Groups, and Universities

The contributors to this volume see an active role for civil society in drawing attention to cases of environmental abuse. Indeed, the rise of the environmental movement in the Caribbean is closely tied to the growing importance of non-governmental organizations (NGOs) in policy dialogues.[4] These nonprofit organizations typically have a full-time staff, a director, a budget, and an office. They range from well-funded organizations with close ties to high-level government officials and international organizations to small, struggling groups of local activists.

Much has been written about NGOs and their capacity to deliver services efficiently, to enhance political participation, and to influence global environmental debates. They rose to prominence in the late 1980s as international donors and bilateral assistance agencies turned to them as promising alternatives to government agencies. The NGOs were viewed as participatory, democratic, and efficient, in contrast to bureaucracies, which were often characterized as corrupt and inefficient. As some of these organizations gained power and prominence, enthusiasm for them was tempered by a certain distrust. Critics argued that NGOs promoted northern rather than southern environmental agendas, and that their commitment to local participation was more rhetorical than real. In fact, the survival of even the most committed of Caribbean environmental NGOs has depended upon funding from transnational NGOs like the Nature Conservancy, Oxfam, and World Wildlife Foundation or from bilateral assistance agencies like USAID, the Japanese International Cooperation Agency (JICA), and the German GTZ. Moreover, there *is* a real difference between NGOs and community groups; the former do not fully represent local concerns.[5] Yet for all their limitations, Caribbean NGOs may be in the best position to carry out biodiversity conservation projects (Jacobeit 1996). They have contributed to more efficient delivery of urban services and have created new (albeit small) spaces for popular participation in environmental and natural resource decision making (Reilly 1995).

Civil society is more than an aggregate of individual NGOs. It is more properly defined as the sum of organized social groups, from soccer clubs to advocacy organizations, but excluding the state and business enterprises.[6] Several authors in this volume express the hope that a regionwide civil society will become part of even larger transnational advocacy networks; they argue that participation in such efforts can significantly enhance the political resources available to Caribbean actors. Caribbean migration to the cities of North America and western Europe has been an important catalyst in the formation of these networks. It can be argued that the shift in emphasis from conservationism to environmental justice in the region is in large part a product of the growing participation of Caribbean migrants in urban environmental movements in the north. Conversely, as Gandy (2002) suggests in his analysis of the environmental justice movement in New York City, the framing of the environmental justice discourse in North American cities—even the definition of urban nature—owes much to the growing influence of the city's Caribbean population. This influence is manifest in struggles over lead contamination, asthma, and garbage and in the ongoing battle for the survival of community gardens in New York City as green islands and performance spaces.

Paradoxically, while Caribbean participation in transnational advocacy networks has increased dramatically with the facilitation of internet communication, environmental NGOs appear to be losing influence. The latter phenomenon may be due to reductions in funding from bilateral assistance agencies, to the aging of the environmental movement, and to the movement of some key NGO leaders into government positions. It may also be due to a growing reliance on municipal governments, universities, local foundations, and the private sector as change agents on the part of international donors and lending agencies like the World Bank and the Inter-American Development Bank, perhaps as part of an increasing programmatic emphasis on governance and decentralization. In the Dominican Republic, for example, environmental programs and projects are increasingly undertaken under university auspices, and with support—and often active participation—from local foundations.[7]

The shifting roles and influence of these different sets of environmental actors have produced changes in Caribbean environmental debates. With the urbanization of island populations, the decline of agriculture as a contributor to foreign exchange, and implementation of large-scale park and reforestation projects, many conservationist goals are being achieved. Here too, however, we see new challenges. For example, a June 2003 article in a the Dominican daily *Listin Diario* charges that lots within protected areas have been illegally sold to wealthy clients for vacation homes. Nonetheless, the major environmental debates in the Caribbean, as in much of the world, now revolve around such urban issues as smog, water quality, and the management of sewage and solid waste.

The Contributors

This volume draws on the experiences of academic social scientists, policy makers, and activists living and working in the United States and the Caribbean. Venezuelan researcher Francine Jácome offers an assessment of efforts to improve the environment throughout the Caribbean region. She finds that most Caribbean governments and NGOs take a local approach to environmental management, thus hampering regional environmental cooperation. She goes on to document some attempts at regional cooperation involving NGOs, government, and international organizations—in particular the United Nations Environmental Program (UNEP). These efforts have emphasized information-sharing and have been undergirded by a growing perception among citizens of Caribbean nations that they share a space as well as a Caribbean identity.

Other contributors focus on particular cases. A common theme in these in-depth analyses is the region's weak position in the global economy. Shaped by external forces—first colonization, slavery, and plantation agriculture, and later contract farming, labor migration, export-platform industrialization, mineral extraction, and tourism—the Caribbean region entered the twenty-first century reeling from the effects of structural adjustment and competing in the globalized economy on disadvantageous terms. Cuba, of course, remains a special case. Efforts to ease the U.S. embargo during the Clinton administration (*New York Times* 2000) have given way to renewed economic and political pressures from the United States with increasing support from Europe. The case studies in this volume address national responses to these global pressures, but suggest that none has been very effective.

In Part Two, a series of case studies examines the impacts of tourism on Caribbean environments. Political scientist Marion Miller documents the commodification of Jamaica's natural and cultural resources for foreign consumption. She argues forcefully that the privatization of natural resources to benefit the tourist economy contributes to continuing social inequality. In particular, she criticizes the seemingly positive decisions by Caribbean states to set aside lands for national parks and protected areas and to promote ecotourism as a strategy for environmental protection. Miller observes that the creation of parks in Jamaica has historically occurred at the expense of local communities and that, since calls for local participation have been largely rhetorical, these communities have come to view ecotourism as a threat rather than a positive development strategy.

Manuel Valdés Pizzini traces the rise of the Puerto Rican environmental movement, placing particular emphasis on struggles over access to the island's coastal areas, which have experienced rapid and largely unplanned development for tourism, residential use, and recreation. Access to the coast has become an environmental justice issue in Martinque as well, as discussed in the contribution by geographer Maurice Burac. Martinique is a department of France, and so—like Puerto Rico—it is politically subordinated to a nation-state with a well-developed legal and regulatory framework and the institutional capacity to enforce environmental regulations. Both the Puerto Rican economy and the economy of Martinique were once based on export agriculture. However, unlike Puerto Rico, whose economy is based on tourism and manufacturing, Martinique's economy is currently based almost completely on tourism. Not surprisingly, as Burac shows, environmental struggles in Martinique from the 1970s to the present have revolved around mass tourism. It could be argued that in these relatively well-off islands, environmental activists can demand a greater degree of accountability from elected officials and more responsiveness from the judiciary. Yet because party politics in Puerto Rico and Martinique continue to revolve around status issues, environmental movements on these islands have often found it difficult to get environmental issues onto local political agendas. Even in this context Valdés Pizzini and Burac see openings for environmental activists: Valdes Pizzini argues that the rapid and inequitable development taking place in Puerto Rico will inevitably prompt a cohesive environmental response, possibly of the type seen around the Vieques issue. This may also be true in Martinique at some future date. According to Burac, however, as of the late 1990s Martinique's environmentalists had generally agreed that they could best further their cause through skillful utilization of the French legal system.

Part Three goes "Behind the Beach" to look at agriculture in the Dominican Republic and Cuba, the environmental consequences of rapid industrialization in Puerto Rico, and the impacts of six decades of military activity on the island of Vieques, Puerto Rico. Sociologist Barbara Lynch focuses on the agricultural sector and its environmental impacts, and compares Cuban and Dominican strategies for agricultural sustainability. She argues that in these countries, and in the Caribbean as a whole, promotion of environmentally destructive export-oriented agriculture has historically come at the expense of domestic food security. Local food producers have not only become economically and socially marginalized, they have been pushed to increasingly marginal lands and have borne much of the blame for environmental degradation. Lynch notes that both Cuba and the Dominican Republic have undergone agrarian reforms, but in neither case did reform produce a nation of small, independent farmers. With the fall of the Soviet Union and a growing food crisis, Cuba out of necessity embarked on an unprecedented effort to achieve agricultural sustainability. The Cuban transition to sustainable agriculture is far from complete, but it has provided a model for other developing countries. Lynch concludes that in both countries agricultural sustainability and food security will become more closely tied in the future.

The "tourist gaze" (Urry 2002) focused on sand and sun occludes the fact that the Antilles are predominantly urban.[8] In several islands, urban environmental issues are assuming a new salience. A chemist, activist, and founding member of Puerto Rico's Green Party, Neftali García has helped to define Puerto Rican environmentalism as an environmental justice movement whose goal is to reverse the negative impacts of Operation Bootstrap industrialization on the island landscape. In this collection, he and his colleagues offer a geography of

Puerto Rican pollution that adds to the growing body of evidence showing the proximity of poor communities and polluting facilities.[9] Their priority issue is water quality and supply, followed in importance by air pollution.

McCaffrey and Baver examine the environmental consequences of six decades of U.S. military training on Vieques. In Vieques, approximately nine thousand residents live with toxic waste as well as large quantities of unexploded ordinance—yet another consequence of the colonial perception that Caribbean human and natural resources exist to meet the needs of states and consumers in the global north.

One objective of the conference that gave rise to this volume was to trace the links between environmental movements in the islands and those coming from the Caribbean diaspora in New York and to better understand the transnational character of the communities that gave rise to these movements. Two chapters in Part Four focus on organization of New York's Caribbean communities in pursuit of environmental justice. In their essays city planner Ricardo Soto Lopez and political scientists Immanuel Ness and Lorraine Minnite, like Harvey (1999), see the struggle for environmental justice as grounded in local experience, but driven by larger struggles for recognition, respect, and empowerment.

Soto Lopez bases his discussion on his experience as a planner and activist in the South Bronx, but he argues that his findings can be generalized to cities throughout the northeastern United States that have large Latino communities. Citing the example of converting industrial land into municipal waste facilities in New York City, he argues that lack of uniformity in land use and environmental protection policies drives the quest for environmental justice in poor communities. He contends further that brownfield revitalization and community participation in planning can enhance economic opportunities in Latino communities while reducing environmental health risks. Soto Lopez concludes with a practical framework for analyzing community land use in poor urban neighborhoods. Ness and Minnite focus on environmental health concerns in the Brooklyn neighborhood of Sunset Park. Like García and Soto Lopez, they draw attention to the ways in which public health and environmental issues overlap in poor urban neighborhoods.

Conclusion

The purpose of this volume is to introduce North American readers to the environmental issues that concern Caribbean people as well as the social scientists and activists working in the Caribbean. For this reason, we have paid less attention to natural disasters or questions of protected area management and biodiversity conservation, and more to topics that fall within the environmental justice domain: *access* to land and other resources and *exposure* to environmental ills. A second objective is to help to dismantle the barriers to interaction among Caribbean peoples that have resulted from differences in culture, language, and governance by drawing out the common threads in the challenges they face whether on the islands or in diaspora communities. If there is one lesson in these essays it is that improving the region's environment and its options for sustainable development will require political will on the part of the state and elites as well as citizen action.[10] Political will is not easy to forge in a region where the costs and benefits of participation in a globalized economy are so unevenly distributed. Still, these essays remind us that a concept of environmental justice is widespread among the peoples of the Caribbean and that even where it is not explicit, it undergirds a great deal of environmental activity.

The contributors to this volume generally agree that the region's environmental problems are not amenable to facile solutions and that quick technical fixes are often unjust. For example, ecotourism projects may meet narrowly defined conservation objectives, but privatization of natural resources and enclosure in support of ecotourism exacerbates problems of inequality and access. Pushed by demands from local organizations and transnational advocacy networks, Caribbean states are increasingly struggling to adopt comprehensive environmental policies that would promote sustainable human development. Our volume offers no simple policy prescriptions. However, a common sense of equity demands that regional actors accord priority to basic environmental and public health concerns. If Caribbean states are to attend to global environmental concerns like biodiversity loss and global warming, they must resolve local problems like water quality, sewage treatment, worker health and safety, and toxic waste management. Poor states cannot easily grapple with both sets of issues simultaneously. However, resolving basic public health problems is likely to have positive implications for climate and biodiversity.

The growing number of region-wide efforts at information dissemination suggests that Caribbean environmentalists have begun to transcend linguistic and cultural barriers to environmental cooperation.[11] Exchange of information about policy and project successes and failures is important, as is the exchange of information on legal and regulatory frameworks and on attempts to achieve policy harmonization within the region. A common regional legal and regulatory framework could do much to prevent the "race to the bottom" in which countries competing for foreign investment sacrifice their natural patrimony and weaken or ignore environmental safeguards. Ultimately, however, there are limits to cooperation. Political will must arise within the individual nations of the region. While transnational alliances have proven useful in getting regional issues onto international agendas, the framing of environmental issues and their positioning in the political arena will of necessity differ in each society. It is up to environmentally concerned citizens on each of the islands to position issues in a way that will capture national attention, and it is up to political elites to face up to the need to improve the quality of life of all citizens.

We conclude then with the leitmotif of this volume—the emphasis on citizen action to promote environmentally sound resource management and development. Without a well-organized and cohesive push from the bottom, change at the top is unlikely. Unfortunately, our studies reveal that much of what passes for community participation is little more than cooptation. Our job as environmental social scientists and activists is to document cases where local participation in the policy process is meaningful and effective, to learn from these cases, and to adapt them to the diverse and constantly changing realities of the Caribbean.

Notes

1. To date, only a few social science studies focus squarely on Caribbean environmental issues. Exceptions are Besson and Momsen (1987); Murray (1991); Barker and McGregor (1995); Collinson (1996); Patullo (1996); McGregor et al. (1998); and Sheller (2003). Sheller's chapter on "Nature, Landscape and the Tropical Tourist Gaze" traces the conceptual development of the Caribbean island as a beach-and-foliage fantasy back to its sixteenth-century origins.
2. See, for example, the discussion of the gendered impacts of Dominican agroforestry programs in Rocheleau and Ross (1995). Roberts and Thanos (2003) treat Latin American environmental crises as expressions of global structural inequalities.
3. For a literary view of the problems caused by mass tourism in the Caribbean see Kincaid (1988). On ecotourism, see Honey (1999) and McLaren (1998).
4. For a discussion of the political roles of NGOs in the Global South, see, for example, McDonald (1997) or Fox and Hernandez (1992).
5. For more nuanced discussions of the roles NGOs have played within the Dominican environmental arena see Sharpe and Lynch (1993), Rocheleau and Ross (1995), and Paniagua (1998).
6. Fisher (1998) offers a general overview of the role of NGOs in international development.
7. A good example is the urban environmental work the Santiago-based Centro de Estudios Urbanos y Regionales has done in poor urban neighborhoods in collaboration with community groups and municipal authorities and with support from a local foundation (Corral 2003).
8. The Caribbean region was considered to be about 70 percent urban in the year 2000 in contrast to 60 percent for the entire Latin American and Caribbean (Heileman 2005: 27).
9. The proximity of polluting facilities to poor populations of color has been a key theme of the environmental justice movement in the United States since publication of the United Church of Christ Commission on Racial Justice's report on Toxic Waste and Race in the United States in 1988. The movement has gained strength and credibility in the past decade (Rosenbaum 2002).
10. Diamond (2005: chapter 11) makes much of this point in his attempt to account for the vastly divergent histories of Haiti and the Dominican Republic. His argument, however, is flawed by an over-reliance on Dominican perspectives on Haitian environmental history and by its emphasis on protected area designation and its scant attention to policies related to highway construction, urban construction, coastal zone development, and the development of agriculture during the same period. Also in arguing that the environmental movement in the Dominican Republic is unparalleled in the global south, Diamond ignores trends in Latin American and Caribbean environmental activism, some of which are discussed in this volume.

11. Examples include the Caribbean Environmental Program (CEP), an initiative developed with guidance from the United Nations Environmental Program, the Comprehensive Resource Inventory and Evaluation System (CRIES), the Caribbean Community Climate Change Centre (CCCCC), the Environment and Sustainable Development Unit of the Organization of Eastern Caribbean States, and a host of more specialized regional institutions treating problems associated with fisheries, health, natural disasters, coral reef deterioration, and ocean dumping.

CHAPTER 2

Environmental Movements in the Caribbean

FRANCINE JÁCOME

Introduction

Environmental movements in the Caribbean region are extremely heterogeneous, primarily because of the large number of actors involved. The main actors are the international, inter-governmental organizations and agencies; the various domestic government agencies; regional, national, and local non-governmental organizations; and transnational organizations (Jácome and Sankatsing 1992). A second reason for this heterogeneity is the different theoretical approaches to environmental problems. In addition to the significant differences in the structures of these organizations, their activities and projects are guided by differing goals.

The purpose of this chapter is to present a comparative study that will facilitate a diagnosis of environmental movements in the Caribbean. Because of the important differences among Caribbean environmental movements, it is appropriate to examine them within the context of different typologies. An initial attempt is made to describe the main factors considered in preparing the typology used in this study. This typology adopts a basically analytical approach to the main characteristics of Caribbean environmental movements. This analysis, in turn, is used as the basis for examining the progress and shortcomings of environmental cooperation in the region, thus enabling us to determine the main prospects of Caribbean environmental movements, and to offer recommendations that could help foster further cooperation on environmental issues.

Typology of Environmental Movements: An Initial Approach

Three main typologies have prevailed in studies of environmental movements: those based on the historical evolution of the movements; those based on theoretical or ideological positions; and those based on practice or actions taken by environmental movements.[1]

The Historical Typology

This typology (García Guadilla 1991) distinguishes three periods in the evolution of environmental movements: (1) the conservationist-naturist stage, which gives way to the (2) political-ideological stage, which is then superseded by a (3) symbolic-cultural stage. In the 1960s and 1970s, environmental movements, particularly those that constituted part of civil society, participated in the struggle to raise living standards and criticized the prevailing developmentalist model but generally had connections to state agencies and political parties. During this second stage, the political-ideological environmental problems were seen as intimately associated with the prevailing model of capitalist development. Environmentalists saw a need to propose an alternative model of development (Garcia Guadilla 1991; Garcia Guadilla and Blauert 1994).

The idea of interdependence influenced the second-stage environmental movement (Dubois 1993). Environmental thinking took into consideration the importance of international trade, financial flows, the need for exchange of ideas about new development strategies, and the growing importance of NGOs. Also taken into consideration was the potential effect environmental factors might have in improving the lot of the poorest sectors in society.

A process of revision and reflection is now coming to a head. As part of this process (stage 3), the role of the NGO is being reconsidered. New networks are discussing "cooperation and the new challenges of development" (Dubois 1993, 147). These networks are also addressing issues of equity and social justice as they relate to environmental themes.

The major drawback of this evolutionary typology is its lack of correspondence with reality. We do not see a linear evolution of environmental movements, but rather the simultaneous coexistence of all three stages or postures.

The Theoretical Typology

Most authors locate their analyses of environmental movements within a broader discussion of NGOs, and propose to analyze these movements according to their neoconservative, liberal-pluralist, or Marxist theoretical orientation (Macdonald 1992). The neoconservative position (Macdonald 1992) sees the proper role of NGOs as promoting a civic conscience, which in turn would lead to the regulation of potential social manifestations that might endanger the stability of political and economic systems. The NGO, according to this view, becomes the guardian of the status quo. As applied to environmental movements, this position is identified with the so-called "technical-managerial" approaches, which emphasize technical solutions to ecological problems and which have decoupled the problem of environment from that of economic development. Proponents of this approach tend to emphasize ecosystem conservation, practically ignoring its relation to social problems (Gudynas 1992). The conservationist-protectionist posture within ecologism, for example, has basically developed remedial and technical policies. For this reason, this approach is generally considered conservative and traditional within today's environmental movements (García Guadilla 1991; González Martínez 1994).

Another variant of this position has been labeled *conservadurismo-dinámico,* which Guimaraes (1992) defines as "that conservationist position whose discourse on change is intended to ensure that change does not happen." As an example, he notes that even though its projects have done violence to the environment the World Bank has become the principal agent for the financial implementation of Agenda 21.

The pluralist-liberal position starts from the premise that society is composed of a number of groups that seek to defend their specific interests, but do not challenge the system as a whole (Macdonald 1992). These groups are not organized along class lines. In the environmental arena, this position characterizes organizations that act as pressure groups—the ecologists and some proponents of sustainable development. In contrast, a more totalizing position calls for the incorporation of economic, political, social, and cultural dimensions into actions oriented toward the solution of environmental problems and for sweeping programs for medium- and long-term social change. This view postulates a social project that incorporates environmental concerns, not in an isolated form, but rather in interrelation with political, economic, cultural, and social objectives. In the more general NGO panorama, this perspective has been labeled "post-Marxist" (Macdonald 1992). It starts from the assumption that economic globalization has made control over state power a non-issue. In this sense, NGOs are moving forward to challenge domination at all levels of society, including civil society. These NGOs see themselves as "organizations for social promotion" (López-Llera, in Macdonald 1992).

In the context of environmental movements, this perspective becomes the "environmentalist perspective," or, as it is known in other settings, the "social ecologist perspective." Given the nature of Latin American and Caribbean realities, it has been suggested that this is the most satisfactory theoretical perspective for research and action on environmental issues (Gudynas and Evia 1991). Also, considering the differences within this perspective, it still presents a total and dynamic vision of environmental problems in which the ecological is tied to the social, the economic, the political, and the cultural. In this sense, it can be defined as "the study of human systems in interaction with their environmental systems." In the environmental system we can distinguish three subsystems: "the human, the built, and the natural" (Guydnas and Evia 1991, 26).

Within this reality, "environmentalism tries to integrate individual rights, traditional values, collective solidarities, economic self-management, and participatory democracy as fundamental aspects of a new world order, open to the advances of modern science and international cooperation" (Leff 1994, 43). The new components of this fraction of the environmental movement are participatory democracy, decentralized management of productive resources, and sustainable development. Environmental cooperation becomes important because the environment as a global *problematique* has given rise to events and tensions that are global in scale.

This perspective has given rise to the idea of "a sustainable life" (World Conservation Union/UNEP/WWF 1991), a concept defined as a new ethic that includes protection of the environment by communities, provision of a national framework for the integration of conservation and development, and promotion of said "sustainable life" by means of a global alliance. Furthermore, it is held that global and shared resources must be subject to international cooperation. In addition, priority is accorded to enhancing the quality of life of the neediest. This can be achieved only through interrelating ecological, economic, political, and social systems and through the intervention of multiple actors, including governments, intergovernmental organizations, NGOs, and the private sector, as well as individuals and communities.

The Praxis-based Typology ← actions

Environmental movements have also been classified according to their actions or praxis (Viola 1992). A first type consists of interest groups, for whom the environmental *problematique* becomes a basis to compete with other groups in the social arena, groups whose bases lie in other *problematiques,* such as gender. Generally, their activities are oriented toward environmental protection and they support, whether implicitly or explicitly, the existing social order.

A second type is closely associated with the birth of green parties, and is linked to the radical sector of environmentalism, which argues that the environmental movement must assume the position of subordinate in the context of the current social crisis. This type of movement questions the current capitalist system, and thus differentiates itself from traditional social movements. It is located within the general context of new social movements (Viola 1992) like pacifism and feminism.

A third type consists of environmental movements as historical movements. For activists in these movements, the present mode of life is not sustainable; it is a product of the predatory attitudes of humanity and a consequence of the unbridled consumption of material goods. According to this tendency, environmental protection will have a braking effect on this model of consumption. It proposes a "globalization of the environmentalist movement" (Viola 1992, 141). With this end in mind, groups embracing these ideas call for the participation of environmental NGOs in a larger, global movement, independent of whatever ideological orientations they may have. They also emphasize the creation of networks with other actors, whether governmental or nongovernmental.

Although different authors have created typologies of environmental movements with distinct parameters, there are points at which the theoretical perspectives coincide. In most cases, this theoretical posture weighs heavily in defining the goals and in implementing activities and projects advanced by the environmental movements. Similarly, the organizational structure adopted by different actors, internally and vis-à-vis other actors, will be defined by their theoretical perspectives. Our premise is that based on the theoretical foundation guiding their activities, projects, and organizational structures, the environmental movements in the Caribbean can be divided into two groups: ecologists and environmentalists. Ecologists stress the need to preserve, conserve, and protect nature; the environmentalists view environmental problems from the standpoint of how they relate to economic, political, and social development.

The second variable in this proposed typology relates to the goals of the various actors, since, as mentioned above, these goals will determine the activities and projects to be carried out by the environmental movements. The movements can be divided into three types according to their goals: those focusing on the physical-natural aspects; those who seek to change the development strategy, which is where most of those advocating sustainable development are grouped; and those intent on fostering economic, political, and sociocultural changes, and whose goal is not just economic development but socially equitable development. If we focus on a third variable, organizational structure, we can identify four major types of movements: vertical-traditional; horizontal; flexible; and those that eschew any formal organizational structure (García Guadilla 1991).

These three variables provide the framework for examining and classifying the various actors in Caribbean environmental movements, as well as a guideline for examining the problem of environmental cooperation. From a historical standpoint, attention was first paid to environmental problems in the Caribbean in the 1950s when some government agencies and NGOs were created. Over the subsequent decades others gradually joined in, adding to the number of actors in the environmental movements. There was a sharp increase in the number of these movements in the 1980s, due mainly to the growing number of NGOs interested in environmental issues. By 1991, in the eastern Caribbean alone there were thirty-five local NGOs—in which membership ranged from 13 to 1,640—and seven regional NGOs (Island Resources Foundation 1991).

The Ecologists. As already mentioned, actors whose theoretical position is based on the idea that the environment must be defined in terms of protecting and conserving nature are generally called "ecologists." Basically, their position is a conservative one aimed at maintaining the status quo. In discussions concerning the effects economic readjustment policies have on the environment, they argue that protection of the natural environment must be taken into account when implementing these policies.

The goals of the ecologists are fundamentally conservationist-protectionist, and their activities and projects are aimed at the physical and natural environment. Within the NGO arena they are considered "conservative," since their actions are geared toward promoting changes that will not cause imbalances in the existing power structures (Macdonald 1992).

Regarding their organizational structures, and specifically, their participation within the various environmental movements, they tend to adhere to a traditional idea of participation (Henry-Wilson 1990; Lewis 1990). Therefore, these nature/conservation actors' relationships, both internal and vis-à-vis other actors, tend to follow a vertical pattern. In this regard the dominant relationship between some northern NGOs and their southern counterparts has been strongly criticized, because it is felt that the northern NGOs are the decision makers, choosing projects to be implemented, while the southerners merely implement them (Jácome and Sankatsing 1992). Although this type of environmental movement has been widely criticized, especially from the viewpoint of the environmental prospects of the south, it is surprising how strong it is within the panorama of Caribbean environmental movements.

The Environmentalists. In contrast to the ecologists whose ideas about nature and conservation have prevailed in the northern countries, environmentalist positions have typically emerged in the southern countries as a result of their specific conditions. Despite the many definitions of "environment," one common element among southern definitions of this term is that they do not refer to nature or to physical space alone. Usually these groups approach environmental problems from the environment-economies viewpoint or from an environment-society viewpoint.

Environment-economics. Speaking in very broad terms, the problem of development is at the heart of discussions about the relationship between environment and economies. There are two basic positions that are not necessarily incompatible, since they are both guided by the premises of "developmentalism." The first position—called "developmentalist" by some authors—is based on the idea that natural resources must be used rationally, protecting the environment for the purpose of national development. Development is viewed as a technical issue, and achieving this goal would lead to the resolution of social and economic problems.

In contrast to this is the sustainable development position. Although there is no consensus regarding the term's meaning, discussions of "sustainability" have made it clear that in the southern countries this term has been revised to include the necessary balance between environmental protection and the war on poverty, the need for economic growth, and the need for more democracy and justice in north-south relations (Marmora 1992). The overall aims of sustainable development advocates are to reduce poverty and to prevent the implementation of nonviable development projects from producing continued overexploitation and destruction of the natural resources.

From the sustainable development perspective, there has been significant criticism about the implementation of structural adjustment policies, largely because they have had extremely negative consequences for the environment (Chantada 1992). These policies have led to a situation where the drive for export-led growth and increased commodity exports and the need to generate foreign exchange, have promoted domestic agricultural policies that have harmed the environment. Policies encouraging expansion of tourism have had a similar effect. These policies have been implemented without any environmental impact studies. Therefore, "the quest for efficiency and competitiveness is carried out at the expense of biological diversity. This has concentrated wealth and spread poverty" (Chantada 1992).

Environmental actors, faced with these negative results and guided by their various approaches to development, do agree on two fundamental goals: the need for a model of economic development—sustainable in some cases—that gives serious consideration to environmental problems, and the importance of international cooperation. In this regard the collective best interest must prevail, consistent with the need to conserve resources and achieve long-term sustainable development. To this end, it is necessary to use appropriate technologies, foster environmental education, and foster "regional activities involving exchanges and the shared use of resources to overcome poverty" (Gudynas and Evia 1991, 173).

Environment-society. In general terms, environmental advocates in this group argue that, in addition to the economy, achieving sustainable use of resources will affect the political and sociocultural components of society. When criticizing the structural adjustment policies that have been implemented in developing countries, they share the view that the main goal should be formulating and implementing equity-based development programs, and that eliminating poverty must be a central component of environmental agendas. Consequently, discussions of the relationship between environment and economies must go beyond their current parameters in order to encompass poverty-related environmental problems. In this regard, some have suggested that the idea of "environmental democracy" be based on a different type of sustainable development, a development strategy "based in the communities and with their direct participation in managing productive resources. The main goals are to achieve equity, balance, and respect for cultural diversity, and . . . a great number of possible futures" (Leff 1994, 55).

Within these objectives, the prevailing position is that the projects and activities should originate at the local level. Projects should be managed by the participants themselves, who then gain knowledge and skills while solving their own problems. As a result of this type of learning, the communities would be in a position to manage both their environment and alternative development strategies. A part of such projects would be to foster "environmental awareness" to guide these organizations and enable them to find the proper balance between meeting their economic and social needs and maintaining a healthy environment.

Similarly, Lewis (1990) has noted that alternatives at the macroeconomic level must be balanced against those that emerge at the local level. Self-management is an important component of this process. Yet the role of communities and their organizations must be to promote and facilitate change, not to become alternatives for political parties, contrary to some proposals being made in view of the shortcomings and growing criticism of traditional parties.

The actors within this type of environmental movement tend to be local, national, or regional NGOs. They generally favor horizontal relations and flexible organizations, both internally and in their relations with other actors. A good example of this type of relationship is the increasing tendency to establish networks of NGOs working on environmental and other social problems, such as indigenous rights, human rights, or women's rights. Environmental NGOs are part of the various kinds of transnational social movement networks that are now being established.

Environmental Movements and Environmental Cooperation in the Caribbean

With few exceptions, movement actors in the Caribbean define themselves as promoters of change at the economic, environmental, and sociocultural levels. Nevertheless, as indicated in the discussion on the objectives guiding activities and projects, in practice the projects implemented by most environmental organizations, but especially the NGOs, are primarily about conservation. The minority engage in activities in which environment is related to economic development and based on sustainable development proposals. NGOs carry out these projects at the local level, and regional organizations promote them at the regional or sub-regional level.

One of our tentative conclusions is that most of the actors in Caribbean environmental movements follow the ecologist tendency, with its emphasis on nature and conservation. A few actors, however, subscribe to the environmentalist school of thought and stress the need for sustainable development. With the exception of the Barbados-based Caribbean Policy Development Centre (CPDC),[2] we found none that stressed the relationship between environment and society, and the need to deal with environmental problems within an economic, political, and sociocultural context.

The goals being pursued by most Caribbean environmental actors are mainly oriented toward conservation and protection of historical monuments, architectural works, or various elements within the ecosystem—generally in the sea or coastal areas. Other environmental groups focus on improving economic conditions of a specific group in society—for example, farmers, young people, women, or fishermen. Actors working toward sustainable development try to foster economic development that is not harmful to the environment (e.g., promoting ecotourism). In general terms, they advocate for social and economic development that goes hand in hand with protecting the ecosystem, especially with a view toward future needs.

These goals are shared by most local NGOs, most of whose activities are aimed at conservation and achieving economic development at the local level. The regional NGOs also focus on conservation and sustainable development, but from a regional standpoint; and, as already mentioned, only one regional NGO was found to have goals linking environment and society.

Most of the international, inter-governmental, and governmental organizations, along with agencies in the region, engage in activities relating to environmental protection and conservation, environmental education, environmental management, research and development of projects involving environmental policy, design and implementation, and monitoring and redrawing of environmental policies. Few of them work in areas linking economic development and the environment. The one exception is the United Nations Environmental Program (UNEP), one of the most important proponents of sustainable development through its Caribbean Environment Program (CEP).

Of the environmental NGOs, the Caribbean Conservation Association (CCA)—an umbrella association of environmental groups from eastern Caribbean states, Venezuela, Puerto Rico, and the United States—focuses on environmental policies and fosters a regional approach to conservation problems. Most of its activities involve preservation and restoration of monuments, historical sites, and buildings of architectural interest; fostering museums; reinforcing local culture; and promoting environmental education and legislation.

The Caribbean Natural Resources Institute (CANARI), based in St. Lucia, is interested mainly in environmental policy, and it emphasizes projects that foster sustainable development and environmental education. It works with communities to create an understanding of the relationship between economies and environment. One outstanding aspect of CANARI's work is that it prefers long-term projects (eight to ten years); most actors support and carry out only short- and medium-term projects. This regional NGO also stresses the need for communities to play a more prominent role in managing their natural resources, by taking part in research, planning, and implementation of environmental policies.

The purposes of the Caribbean Policy Development Centre (CPDC) are similar, but broader in scope. Projects and policies carried out by this organization aim more at fostering discussion of economic, sociocultural, and ecological development with an emphasis on equity. The CPDC assigns priority to problems of inequality due to ethnicity, race, class, and gender, among other factors.

Despite our limited information, our tentative conclusion is that the activities of the local NGOs are primarily aimed at the protection and conservation of ecosystems, preservation of historical and architectural monuments, development of museums, clean-ups and recycling, and environmental education. Additionally, they foster projects aimed at sustainable development, basically in the areas of fisheries, agriculture, and forestry activities. In a few cases, projects are aimed at developing tourism and related areas. Another type of project, both regional and local, that is Caribbean-context-specific, involves preparedness for natural disasters—mainly hurricanes and tropical storms. Most disaster preparedness activities are limited to the local or national level, and there are virtually no projects designed to establish cooperative ties with other national NGOs to work at the regional level.

It should be recalled that governments, environmental NGOs, and foundations from the northern countries or international organizations—in which funding from northern countries plays an important role—provide most of the funding for the environmental movements in the Caribbean. It should, therefore, be expected that there is a certain degree of influence on the goals and projects of both regional and local NGOs.

Regarding the organizational structure of environmental policy making, at the government agency level the traditional vertical structures prevail. Decisions are made at the ministerial level, with directors of departments or divisions as the next level. This type of structure was also common in intergovernmental organizations. Basically, decisions are made at meetings attended by representatives of the member states and implementation is entrusted to an executive director or the like. This is the type of structure found in most of the regional NGOs, such as CANARI, where the Board of Directors meets twice a year to discuss, evaluate, and plan the organization's activities. Within the CCA, decision making is in the hands of the Board of Directors and the Executive Director. The CCA is a special case since its members are governments, non-governmental organizations, and individuals, the latter being the majority. On the other hand, the structure of the CPDC is based on an Annual Assembly and a Steering Committee that is in charge of planning and carrying out activities.

Regarding staff, the general finding was that most consist of executive, administrative, technical, and consulting personnel. There is little research and academic staff, except in the case of the Eastern Caribbean Center and its program involving the Consortium of Caribbean Universities for Natural Resource Management, a group in which countries from the four Caribbean sub-regions participate.

Given the predominance of the conservationist viewpoint—and its fundamentally local goals—environmental cooperation is difficult to achieve. Environmental proposals focusing on sustainable development also tend to favor local solutions. Thus, our preliminary analysis of environmental movements suggests that a group's theoretical views, which bear on the definition of goals and institutional organization, may become obstacles to Caribbean environmental cooperation.

In general terms, environmental cooperation has been guided by two basic approaches (Serbin 1992). The first emphasizes technical assistance aimed at maintaining the status quo and implementing conservationist policies. The second, espoused, for example, by the Brazilian, Guimaraes (1992), is the "ecopolitics" approach, in which environmental problems are considered as part of the framework of the imbalances between north and south, and the environment is a political issue. The first approach is the one most commonly found in the Caribbean.

An analysis of regional environmental cooperation to date shows that there have been three major mechanisms for regional cooperation (Ragster and Gardner 1993). Government-to-government cooperation, either bilateral or multilateral, has generally been limited to defining policies, although there have been a few cases of help in program implementation. This type of cooperation is fostered by the Organization of American States, the Canadian International Development Agency (CIDA), and the United States Agency for International Development (USAID).

There have also been some cases of cooperation between government agencies and environmental NGOs, international organizations and NGO networks, international organizations and local NGOs, and national governments and local NGOs. In general, this cooperation has involved programs as well as specific projects. The two most important Caribbean programs of this sort are UNEP's CEP program and the programs of the CCA. Last are NGO-NGO coalitions in the Caribbean; both bilateral and multilateral NGO networks have been increasing significantly since the 1980s. Since they work basically on local problems, to date the local NGOs have not had a view toward exchanging experiences or setting up joint projects. This may change, though, as most of the projects being promoted by intergovernmental groups—U.N. projects, for instance—are at the local level. What can hinder information flow within countries is that most of the governments do not belong to networks and that their most important relations are with international organizations; in some cases they are not even in contact with the NGOs in their own countries.

At the regional level, most cooperation occurs among regional NGOs. This is the case with the CCA and CANARI, two organizations with a number of joint projects, both local and regional There is also cooperation between regional and international organizations: intergovernmental organizations such as UNEP, FAO, OAS, and CARICOM, and NGOs from the United States and Canada.

The tentative conclusion on environmental cooperation is that the relationship generally comes about as a function of financial or technical assistance from regional and international organizations to those working at the local level. Regional and international organizations have done little to establish relations among national actors, and they prefer bilateral relations between these organizations, between a local NGO and a regional one, and between a regional intergovernmental organization and a government. No strong evidence was found of any intent to foster interrelationships or multilateral relations among local entities through intergovernmental organizations or regional NGOs.

Furthermore, intergovernmental actors and regional NGOs tend to be circumscribed within a single sub-region. One example of this sub-regionalization is the Caribbean Environmental Health Institute (CEHI), which is limited in scope to the CARICOM member countries. The outstanding exceptions are the CPDC, which has member countries from the four sub-regions, and the CCA, which advises both regional and national organizations and fosters the development of regional projects.

Until now, the mechanisms suggested and used to promote regional cooperation have been organizing workshops and meetings; creating pressure groups; developing networks; and building coalitions by such regional NGOs as the CPDC. Another strategy has been holding congresses, such as the Annual

Conservation Congress organized by the CCA, which also publishes a newsletter and has developed a radio series. Sharing television programs and videotapes has also been suggested as a possible tool. Consulting missions and technical assistance to other countries in the region, participation in regional organizations—either NGOs or intergovernmental organizations—and participating in project implementation have been the most commonly used mechanisms at the government level. And shared advisory and training programs and environmental education programs have also been tried.

Yet, despite the existence of conceptions of regional environmental cooperation and of both specific and general mechanisms for implementing it, we find, especially within the local NGOs, whose work is rooted in local problems, little evidence of cooperative relations that might facilitate exchange of common experiences or formulation of joint projects. Similarly, the majority of projects promoted by intergovernmental organizations—for example, those of the United Nations—are oriented toward the local level. Yet, most Caribbean governments privilege their relationships with international organizations. As a result, we typically find an absence of mechanisms for cooperation with NGOs within a single country.

We conclude, then, that environmental cooperation exists primarily as technical and financial assistance flowing from regional and international organizations to actors working at the local level, with very little mediation by these organizations in relations among national environmental actors. In addition, it appears that bilateral relations—between a local NGO and a regional NGO, or a regional intergovernmental actor and a government, for example—are privileged. In this sense, the principal obstacle to regional environmental cooperation is the lack of awareness of the need to develop relationships among actors within the environmental movements. Such relations should be developed through intergovernmental organizations or regional NGOs, thus overcoming their present reluctance to view themselves as facilitators working with different actors to resolve common environmental problems.

Another obstacle to regional environmental cooperation is that intergovernmental actors and regional NGOs tend to be restricted to particular sub-regions. An expression of this kind of sub-regionalization is the Caribbean Environmental Health Institute (CEHI), whose radius of action is limited to countries belonging to CARICOM. On the other hand, we find exceptions like CPDC, whose members come from the four sub-regions, and CCA, which provides technical assistance to regional and national organizations and promotes development of regional projects.

To summarize, one of the most important factors to consider in regional environmental cooperation is the role regional and international bodies can perform. It is precisely within these organizations that the best opportunities exist to encounter different local environmental experiences and needs; they also present the best arenas for discussion and identification of common problems.

Conclusion

The idea of the Caribbean as a region (Serbin 1996) is fundamental if we are to overcome the barriers imposed by localism, insularity, and sub-regionalization. These are barriers not only to environmental cooperation, but to regional cooperation in general. In this sense, the Caribbean Sea, as a "shared regional patrimony," could become a focal point for environmental cooperation—for joint activities based on environmental stewardship and ecology. This, in turn, suggests the importance of a coordinated effort involving multiple actors, including regional and international organizations, governments, and NGOs.

One of the pillars of the new integration processes in the Caribbean region has been the adoption of the Caribbean Sea as a common good, specifically by recognizing the region's economic potential, common sociopolitical interests, and the similar nature of the ecological and security threats each territory faces (Serbin 1994). Several important proposals exist to overcome the obstacles posed by the lack of coordination among different actors in the environmental movements. For example, the Association of Caribbean States (ACS) includes among its areas of endeavor coordination and collaboration in "processes of foreign policy and international economic relations in regional, hemispheric, international and multilateral organizations" (Lewis 1994, 4), Thus, the ACS can play a role as coordinator of diverse activities in the environmental area as well as in others.

Additionally, the principal actors in organizations for international cooperation are generally governments. But with the matted growth of transnational relations, it will be necessary to take into account the growing pressure of groups in civil society (Serbin 1994). To accomplish this, we must seek mechanisms that permit multilateral cooperation among traditional, new, and emerging social and political actors.

Globalization has led to growing regionalization and transnationalization of relations (Serbin 1994). In this regard the changes taking place in the Caribbean—not only among governmental, intergovernmental, and international actors, but also among the NGOs—must be analyzed from the standpoint of environmental cooperation. The above discussion highlights the importance of the idea of the Caribbean as a region as well as the search for mechanisms by which non-governmental actors can participate in cooperative efforts to confront globalization.

For the future prospects of the environmental movements, joint action among the agencies that formulate and implement policy and the various NGOs (especially local ones) is absolutely necessary. These actors may come together around common principles, objectives, and goals that could lead to joint projects and activities. These relations could be established either for very circumstance-specific short-term projects, or for longer-term programs. Such relations would in no way affect the content of the environmental movements, although they still would depend on the social, cultural, geographic, political, and economic contexts from which they emerged. Caribbean countries share a number of features that make it possible to draw up regional environmental cooperation projects.

Environmental cooperation must be multilateral; thus it will be necessary to overcome old ideas of national interest and strategic security. We must foster the kind of cooperation that will strengthen regional interests while respecting the economic, political, and sociocultural characteristics of the various nations. Indeed, progress has been made in the discussion of environmental problems and sovereignty, since it is increasingly clear that environmental deterioration is an issue of "international interest." The important role NGOs play in dealing with common environmental problems and in world negotiations has also been acknowledged (Hurrell 1992). Essentially, what is needed is a balance between sovereignty and international responsibility.

Serbin (1992) has noted that environmental cooperation, in the framework of regional security, must be guided by four major premises. First is the need to foster a dialogue in which the different actors have the political desire and will to discuss paradigms, hypotheses, and outlooks. Second is the need for a joint approach to topics such as the transfer of clean technologies, harmonization of environmental legislation, assessment of the impacts of resource use on GNIP, development of environmental protection and management policies, and environmental education. Third is the need to move beyond the narrow viewpoint of national interest and delve further into shared interests. Fourth is the need to identify threats to environmental security and common weak points, in order to design common policies. For this reason, environmental cooperation must be multilateral. It will also be necessary to discard old conceptions of national interest and strategic security and to develop new forms of cooperation that strengthen regional interest, while respecting the particular economic, political, and sociocultural features of each nation.

There have been advances in the discussion of environmental protection and sovereignty that conclude environmental degradation is a basic element of "the national interest." These discussions have also recognized NGOs as important actors in confirming common environmental problems, and have foregrounded their role in international negotiations. In essence, the need to maintain a balance between sovereignty and international responsibility is now well recognized (Hurrell 1992).

The fundamental challenge environmental movements face as they seek to promote and develop regional environmental cooperation is the process of regionalization and the incorporation of new social actors. For this reason, recognizing the Caribbean Sea as a shared space where the common good takes priority is a crucial first step in developing a Caribbean identity—an identity that can serve as the basis for cooperation among multiple actors to address common problems.

Notes

This chapter builds on "Los movimientos ambientales y la cooperación ambiental en el Caribe: Una primera aproximación," a paper presented at a 1995 International Seminar, "Hacia una agenda sociopolítica de la integración en el Caribe," hosted by FLACSO Santo Domingo.

1. There were three main sources of data for this article. Some data were gathered from a questionnaire that was part of a larger research project coordinated by INVESP and titled, "Los problemas ambientales en el Caribe: El impacto de los factors socio-culturales" (Environmental Problems in the Caribbean: The Impact of Socio-cultural Factors). Other data came from documentary sources listed in the bibliography. Finally, some information came from primary materials from the following organizations: Caribbean Environmental Health Institute (CEHI), Caribbean Natural Resources Institute (CANARI), Caribbean Conservation Association (CCA), and Caribbean Policy Development Centre (CPDC).

2. The CPDC is a coalition of Caribbean nongovernmental organizations. It lobbies regional and international governments on behalf of social groups in the region whose interests are often ignored in policy debates. In particular, it has argued against structural adjustment programs and for debt forgiveness and increased social spending in the region.

PART II

*The Political Ecology
of Sun and Sand*

CHAPTER 3

Paradise Sold,
Paradise Lost

JAMAICA'S ENVIRONMENT AND CULTURE
IN THE TOURISM MARKETPLACE

MARIAN A. L. MILLER

Introduction

Tourism is the leading trade sector for many Caribbean states, with several of them dependent on the industry for more than 50 percent of their gross domestic product (GDP) (McElroy and Klaus 1991, 144). Although Jamaica has a more diversified economy than some of its Caribbean neighbors, tourism still accounts for about 25 percent of its GDP (McElroy and de Albuquerque 1991, 122).

Jamaica's tourism is based primarily on natural resources like sun, sand, and sea; however, over time, the tourism product has been modified to include cultural elements. These resources have been packaged and sold as "paradise." In an economic context that places a premium on growth, tourism is seen as one of the few areas in which growth is possible. As a consequence, natural and cultural resources are being commodified. This chapter examines the commodification of these resources and assesses the consequences of that commodification for nature, culture, and society. Generally, commodification has contributed to the degradation of both cultural and natural resources. In many cases, cultural practices and objects have been routinized into products for sale to the tourist, and natural resources have been enclosed or privatized in order to ensure their continued availability for the tourism market. In a country where there is significant inequality, the continuing enclosure limits the social and economic participation of a large part of the population.

The Global Context

Tourism strategy in Jamaica and the rest of the Caribbean is largely determined by the global economy. The globalization of economic relationships has made it difficult for individual governments to intervene and manage their economies. But this problem is exacerbated for Third World countries that often have only minimal control over the disposition of their resources. Consequently, the policies employed in a country like Jamaica are often a response to external signals rather than a response to the demands of the local community. Outside interests, such as foreign corporations and consumers in other countries, drive tourism policy. While these foreign and international interests may also be important in developed countries, the size and precarious health of many developing country economies make them more vulnerable to outside interests.

Tourism is perceived as a means of boosting foreign exchange and providing jobs. But in practice, the benefits fall short of expectations. Earnings are often expatriated to foreign airlines or hotel management companies, and the demands of pampered tourists strain limited resources (Miller 1992, 301). Some local entrepreneurs have become involved in hotel development, but many hotels are still foreign-owned. The industry is also very dependent on imports to cater to the foreign clientele, so this increases the outflow of currency. In addition, the negative impacts on the environment are everywhere evident.

In designing the local tourism product, decisions about resource use are made with the interests of foreign consumers in mind. In developed countries, the tourism product is targeted at both local and foreign consumers. However, in a small developing country like Jamaica, few locals can afford to purchase the tourism product (Miller 1992, 301), and tourism developers cater primarily to the external market. This, of course, affects how resources such as nature and culture are presented and consumed as parts of the tourism package.

Tourism and the Commodification of the Natural Environment

The Jamaica tourism industry had its beginning in the late 1800s in conjunction with the development of the banana trade with the United States. Banana boats ferried wealthy Americans into the country, and banana companies were pioneers in providing hotel accommodations (Taylor 1993, 37). The United Fruit Company dominated the first few decades of the tourist industry (Taylor 1993, 137). From the early days, the island's natural beauty was its major drawing card. Before 1914, Jamaica's highland areas were big attractions, but after World War I, sea and sunbathing became significant draws (Taylor 1993, 140). As the tourism industry developed, it became more and more focused around coastal areas. These areas include coral reefs, seagrass beds, mangrove forests, and coastal lagoons. They are ecologically important because they are vital to the maintenance of marine life and to the stability of coastal features. But they are also economically important because of their centrality to activities like fishing and tourism. More recently, attention has turned again to the island's interior. Although not as popular as the coastal area, the mountainous interior is once more becoming an important part of the country's tourism product.

Because tourism depends so much on the exploitation of natural resources, tourism operators want access to and control of these resources. They try to ensure this by buying or leasing them. Where this has not been possible, they have actively encouraged of the state to enclose resources in national parks. Sometimes this is a strategy to slow the degradation of resources. Because tourism depends on an apparently clean environment, there is a concern to limit use and pollution of certain resources. This control may be achieved by controlling access. Because of the primacy of tourism and tourist interests, tourists are not the ones whose access to resources is limited. So, although these are ostensibly public parks, they are clearly targeted at the tourism market.

Tourism degrades the environment by increasing the demands placed upon the capacity of the area to assimilate wastes, as well as by dredging harbors and by building hotels, marinas, and resort areas. Hotels are prime sources of water contamination. Development of marinas and harbor facilities generally adds to problems of pollution, such as human waste disposal, destruction of mangroves, coastal siltation, and oil leaks from engines. Tourism and other recreational activities also increase damage to coral reefs and grass bottoms. Physical damage to coral reefs is caused by extensive yacht anchorings and coral harvesting. Recreational uses such as boating add to the accumulation of plastics and other trash in the near-shore and coastal areas.

Because of continuing environmental damage—as well as the need to extend the consumer–base—ecotourism, the new trend in global tourism, has been strongly promoted. This approach places a premium on the establishment of protected areas, such as marine parks. Ecotourism is an amorphous concept and it has a range of definitions.[1] However, the Caribbean Tourism Organization has formulated a definition, which it has recommended to the region as the official meaning of "ecotourism": "The interaction between a visitor and the natural or cultural environment that results in a learning experience, while maintaining respect for the environment and culture and providing benefits to the local economy" (Brereton 1993).

An examination of this definition reveals three key components: the interaction between the visitor and the environment, respect for that environment, and benefits for the local economy. These were also relevant factors in the days before ecotourism, and they inform the conflict over natural resources. But this definition does not reflect concerns about the distribution of benefits and issues of equity, factors that are relevant in stakeholders' conflicts about the disposition of natural resources. Although ecotourism is supposed to have a strong local component, ecotourism projects are usually externally planned and managed, and the infrastructure is often foreign-owned. There has been limited effort to support forms of ecotourism that are locally initiated and managed and that support community development objectives. Instead, it is assumed that some economic benefits will trickle down to the local population through employment or new developments put in place to enhance visitor comfort and enjoyment. Weighed against those limited benefits are the sacrifices and trade-offs required of communities in the affected area. Subsistence resources, including firewood, grazing lands, and fish, can be closed off and earmarked for tourism, resulting in a net loss for the local population. Although ecotourism proponents stress the centrality of "local participation," in actual practice the projects are centrally or externally controlled. Local people are consulted perfunctorily and they provide a labor force, but in most cases that is the extent of their participation. And very often the consultation is done primarily as a means of seeking local support or neutralizing local resistance. So, rather than being a means of local development, ecotourism can be perceived by local communities as a threat to their livelihoods and way of life. In some cases, protected areas have become the battleground between ecotourism and rural communities. And in only a few cases is there a significant effort to realize the rhetoric about local participation. Generally, ecotourism has increased the trend toward privatization and other kinds of enclosure of resources.

Debate about the appropriation of natural resources for tourism addresses not only tourism strategy, but also the environmental consequences of the proposed strategy. A variety of stakeholders participate in this debate. Stakeholders with interests in the disposition of Jamaica's natural resources include the government, developers, tourists, and tourism industry workers. But since the industry involves multiple-use areas, other people are affected by the appropriation of resources for tourism. These include coastal dwellers, artisanal fishermen, and small farmers. Sometimes these actors find that their interests diverge from those of tourism developers.

Coastal dwellers are both winners and losers as a result of mass-based tourism. Their employment might be a direct or indirect result of tourism, but they also experience decreased access to beaches, rivers, and falls; they share a degraded environment with visitors; and their residential choices are limited by inflated real estate prices. Both fishermen and small farmers depend on access to a certain share of natural resources. But with the growth of ecotourism, they are seeing their occupational interests come into conflict with tourism interests. Of the affected stakeholders, the tourism entrepreneurs are in the best position to influence government decision makers. They move in the same social circles, and their access is enhanced by their involvement in an enterprise that is regarded as crucial to the region's economy. Other stakeholder groups, such as the artisanal fishermen, coastal dwellers, and small entrepreneurs have few avenues of influence. A brief look at tourism development in the near-shore marine areas, the coastal zone, and the mountains will illustrate how the various stakeholders' interest have been served.

Near-shore Marine Areas. One approach to dealing with the degradation of these areas has been the establishment of marine parks. One example is the Montego Bay Marine Park. Opened in July 1992, this park is divided into four zones, two of which are classified as replenishment zones where fishing is limited. Water sports are allowed in the other zones (*Jamaica Weekly Gleaner* 1992).

The park regulates the activities of two major groups of actors: tourism operators and artisanal fishermen. It has been easier to communicate with the tourism operators since they recognize an immediate interest in the success of the park. In contrast, many fishermen are concerned with their own individual interests, and are not organized to pursue their group interests (Hall 1995). Exclusion from the park has been a cause for resentment among fishermen, since fishing grounds that are off limits to them are available for tourist activities. They are angry because they now have to go to more distant fishing grounds. Some fishermen have violated the new regulations, and this has resulted in arrests and convictions.

The Coastal Zone. Jamaica's beaches and coastal wetlands are also contested areas. Jamaica has 488 miles of coastline, of which only nineteen miles remained public in 1992. By law, the land below the high tide watermark belongs to the public (Taylor 1993, 170), but much of this beach property has been leased to tourism developers or claimed as private property. Some developers have paid as little as $10 per year for the right to use particular stretches of beach (interview with John Maxwell, June 18, 1992).[2] As a result of this enclosure process, many Jamaicans have been effectively excluded from the island's best beaches.

Conflicts also occur over wetland resources when protection for ecotourism takes precedence over resource exploitation. This is especially a concern when subsistence uses are displaced by tourism. A wide range of artisanal occupations and crafts can be found in the wetlands: for example, in the Black River Morass, there is fishing, shrimping, thatch cutting, and traditional handicrafts. And sometimes, disagreements arise within the tourism industry over the most appropriate use of coastal land. In 1992, controversy erupted over a proposed hotel development on a twelve-acre seasonal wetland. Opponents within the industry claimed that it was the last remaining area of standing wood in Negril (Ansine 1992) and they wanted it developed as a nature park. In the end, the hotel developers got the go-ahead from the relevant authority.

Mountain Areas. The ecotourism strategy has led to the commodification of mountain areas previously excluded from the tourism product. A major consequence of this change was the establishment of the 78,000-hectare (approx. 193,000-acre) Blue and John Crow Mountains National Park in 1993. But the park was not uninhabited. It was home to a number of small communities that depended on its water, soil, and forest resources. Resident farmers regarded the thousands of acres of forests as their lifeline; they cleared the land to plant crops or cut trees to sell. When the park was established, those farming the protected area had to be removed. Many farmers saw the creation of the park as victimization, with land being taken away from them and given to the "big men." Although there was a relocation process, they regarded the new land as marginal. And the effort to involve the people living on the fringes of the park in the ecotourism project was not comprehensive (Neufville 1993, 16). Consequently, these people feel no sense of "ownership" of the protected resources, which are perceived as being protected for tourist consumers and tourism entrepreneurs, but not for locals.

There are just a few examples of real community participation in ecotourism in Jamaica. One of these is Top of Jamaica Blue Mountain Tours (TOJ). This twenty-five-member tour company grew out of one of the Local Advisory Committees formed by the management of the national park as part of its community involvement effort. TOJ was established to manage hikers' cabins, to provide guides for visitors to the Park, and to provide other services in ways supportive of the Park's conservation objectives. This case addresses some of the need for economic alternatives to cultivating the forest (Levy 1987, 8).

Tourism and the Commodification of Culture

In the 1970s, one writer pointed out that it took "a particular history to accept that the external manifestations of one's culture are valuable chiefly as ornamentation for hotels designed, constructed, and managed in the interests of overseas profit" (Hiller 1979, 28). Jamaica, a place where no family's roots are more than a few centuries deep, may be seen to have such a history.

Jamaica' s history and culture have been shaped by a number of factors, including its African heritage, European colonization, slavery, indentured servitude, and the struggle for freedom. European colonization rewrote the culture of Jamaica by completely wiping out the indigenous population and replacing them with African slaves. Later on, indentured laborers from Asia were added to the mix. All of these factors have been reflected in the evolving forms of music, theater, literature, and visual arts. For the young nation, the challenge has been to build a coherent cultural identity out of these diverse strands. Given tourism's dominance, the challenge is to prevent culture and people from being transformed into mere commodities for the tourism industry and reduced to providing service or local color for holiday pictures.

From its early days, the tourism industry has been criticized for restoring elements of the plantocracy. It has been regarded as a new cash crop, replacing the dying sugar industry. Over the decades, critics have pointed to an industry employment structure that has perpetuated the colonial class division where white foreigners control and manage, while local black people work and serve (Conway 1993, 170–171). This picture has changed somewhat in the past two decades, as Jamaicans own and manage a larger share of the tourism accommodations, but the bulk of the tourists are still white, and those who serve them are usually black. For the most part, industry promoters have portrayed Jamaica and other Caribbean destinations as a paradise of beaches and flowers, with smiling black people serving a totally white clientele. This one-dimensional picture ignores the social, ethnic, and occupational diversity present in Jamaica (Holder 1993, 25).

The demands on the entire population to please the tourist are a much more important factor in a small, developing country like Jamaica than in an industrial country like the United States or Britain, where it is recognized that people have concerns and occupations unrelated to the tourism industry. Tourists in these countries are not particularly deterred by curt Customs and Immigration personnel. In contrast, in Caribbean destinations like Jamaica, the whole population is charged with being nice to the tourists (Patullo 1996, 25). Local people are seen as important components of the tourism product, so there is pressure on them for more than competence. It is not enough for customs agents and the immigration officials to work expeditiously; they need to smile as they do so.

There is also concern about cultural dependency as Jamaica's culture is increasingly conditioned by its exposure to U.S. and other foreign influences. In the brief period since independence, much nation-building has taken place, some of it expressed in drama, visual arts, music, and literature. But the shadow of dependency remains. Much of what is admired is "foreign," and so often those preparing the cultural components of the tourism products are not concerned with protecting the authenticity of art and cultural forms.

Industry packagers tend to caricature Jamaican culture as limbo dancing, fire-eating, steel-band music, streetside wood carvings, and friendly Rastafarians. This has led some to criticize the tourism industry as a "modern plantation that is destroying and degrading what is unique in the country." Elements of folk culture have become marketable commodities, readily and monotonously packaged as tourist entertainment. Yet the package offers only a watered-down picture of the cultural offerings of the country.

Music has always been an important part of the Jamaican tourism package. But too much effort has been spent on the presentation of hackneyed songs like "Yellow Bird," when there is a wealth of folk material that could be shared with visitors. Jamaica's musical creativity has also been evident in the development of original genres of popular music, such as ska, rock-steady, and reggae. The latter has gained international attention through the work of artists such as Bob Marley, Toots Hibbert, and Jimmy Cliff. In its early days, reggae was primarily a music of protest. The protest element was modified as it sought greater international acceptance. Reggae is now an important component of the tourism package. The music festivals, Reggae Sunsplash and Reggae Sumfest, were launched primarily as tourism products. However, they are rooted in the community and have not been entirely handed over to tourists. Jamaicans make up the majority of the attendees at these festivals.

Jamaican dance has fared less well under tourism commodification. The packaging has distorted perceptions of Jamaican dance theater. Based on the entertainment packages provided to them, tourists would believe that Jamaican dance theater consists primarily of limbo dancing and fire-eating. But if they were to venture beyond the tourist enclaves, they would find elements of a Caribbean dance theater with a distinctive style and content. It draws on African memory, as well as European influence and masquerades. On this basis, the National Dance Theater Company of Jamaica has established an extensive repertoire of dance-dramas and pure dance compositions (Nettleford 1985, 29). Their performances provide a more authentic representation of Jamaican culture than does either the limbo dancer or the fire-eater.

Commodification is also rampant in the area of "tourist art." Jamaica has a thriving art movement, and its creative artists work in a variety of forms. But many artists and craftsmen have modified their work to produce the kind of stereotypical products they believe tourists want. At one time, preferred stereotypes were pictures of coconut trees and sunsets, or market women with baskets on their heads; now one of the stereotypical images is the dreadlocked Rastafarian. That is the image tourists are supposed to want to take back home to share with their friends, so carvers and painters turn out hundreds of these hackneyed images. But at least this particular souvenir is made in Jamaica. Many of the souvenir items for sale in the tourist enclave shops may have the word "Jamaica" stamped on them, or woven into them, but they are not produced on the island.

The influence of tourism can also be seen in the renewed local attention to history and heritage. Some regard this as a positive development, since it involves the restoration and maintenance of historic buildings. The Seville Great House and Heritage Park, opened in 1994, represents one example of official efforts to harness heritage tourism. Seville is in St. Ann's Bay, an area with a rich and varied history. It was once an Arawak settlement, then the first Spanish capital of Jamaica, and later, under British occupation, a sugar plantation with a Great House and a village of African slaves. St. Ann's Bay was also the birthplace of Marcus Garvey. At the opening ceremony for the project, the Minister of Education and Culture emphasized its value for Jamaicans as well as tourists. But note that in this case attention to heritage was spurred by falling arrival rates in the older tourism destinations like St. Ann's Bay; it was not response to local needs (Patullo 1996, 198). The Tourism Action Plan's agenda includes the improvement of towns and villages, to recognize the contribution of Jamaican craftspeople to the creation of Jamaica-Georgian architecture (Patullo 1996, 190).

Culturally, all has not been flattened and routinized. Important parts of the culture remain outside the packaged paradise. Traditions of religious rites and bush medicine remain outside the prepared package, and they flourish in the rural areas and on urban streets where tourists never go (Patullo 1996, 198). Although religion in Jamaica is dominated by Christian elements from Europe and the United States, some African religious forms have survived the move across the Atlantic. The influence of this is seen in *Pukkumina (Pocomania)*. Bush medicine has also survived the arrival of conventional medicine. Knowledge of special plants has been passed down for generations, and local herbs are used for tonics and medicine. For example, the ganja plant[3] is prized for its tonic tea and herbal extract. Because of the enclave nature of mass tourism, there are still areas not yet incorporated into the tourism package, not yet reduced to T-shirt art.

Conclusions

At a fundamental level, nature and culture are community resources. But with their increasing commodification, most citizens have very little input into their use and management. This places increasing strain on a society already riven with inequities. In spite of the enclosure and degradation of these resources, there is no organized movement to dismantle tourism. Rather, the focus is on managing the resources more efficiently. However, the industry does have its critics. Some see it as a new plantocracy, replicating the economic and social relations of slavery. They regard the current mass tourism strategy as environmentally, socially, and culturally unsustainable. One option put forward is a move from mass tourism to niche tourism areas, such as nature tourism or retirement tourism. These strategies would be less damaging to natural and cultural resources. And they might allow local communities more effective input into the management of community resources. However, a detour to either of these niches is unlikely to happen soon; developers have too large an investment in the selling of paradise to the mass tourism market—an unsustainable path that will eventually mean the loss of "paradise" to both locals and tourists (who will simply select alternate destinations). Already, to paraphrase a calypso, some locals complain that because of tourism, they feel like aliens in their own land.[4]

Notes

1. John Junor, Jamaica's Minister of Tourism and Environment, described ecotourism as "a marriage between development and conservation," in "Ecotourism Vital, Says Junor," *Jamaica Weekly Gleaner*, July 10, 1992, p. 11.
2. Maxwell was chairman of the Natural Resources Conservation Authority and related authorities in 1977. The interview was conducted at CARIMAC, University of the West Indies. Transcript in author's possession.
3. This plant is also called marijuana.
4. This sentiment was expressed by Rohan Seon in the calypso "Alien."

CHAPTER 4

Historical Contentions and Future Trends in the Coastal Zones

THE ENVIRONMENTAL MOVEMENT IN PUERTO RICO

MANUEL VALDÉS PIZZINI

Introduction

A "new" space for leisure and a landscape of high aesthetic value, the coastal zone of Puerto Rico attracts numerous visitors and investors.[1] With commodification, coastal lands are attracting well-to-do home buyers who are displacing long-term local residents of rural coastal communities, fishing villages, and small harbors. Real estate development has focused on construction of condos, resorts, and houses, sold at prices that make them unaffordable for a local population suffering from unemployment and poverty. Having observed the growth of recreational activities and infrastructure on the coast (Valdés Pizzini, Chaparro, and Gutiérrez 1991), in 1992 my colleague Jaime Gutiérrez Sánchez and I embarked on an investigation of this transformation and the associated displacement of traditional settlers. One aspect of this investigation was a study of local responses to the environmental impacts of coastal development.

This process of "coastal gentrification" entails new architecture, social practices, lifestyles, languages, and ethnic groups (Smith 1996, 41; Iranzo 1996). We assumed that gentrification, along with development and urbanization, was highly correlated with an increase in community-based and environmental organizations. These organizations were formed to protect nature and to preserve the social and cultural integrity of local communities threatened by economic change.

Our previous research on fishing communities gave us a relatively thorough understanding of socioeconomic and landscape transformations in the western municipality of Cabo Rojo (see Valdés Pizzini, Gutiérrez Sánchez, and González, forthcoming). Building on the Cabo Rojo case, we designed an exploratory project in the municipalities of Guánica and Lajas in the southwest of the island, and in Aguada and Rincón in the west.[2]

Drawing on the results of this project, this chapter addresses the environmental problems occurring in Puerto Rico's coastal zone and offers a historical perspective on the environmental movements that developed in response to these problems.[3] I argue that unsustainable growth in the coastal zone is to be expected. Therefore, civil society will continue to play a key role in the stewardship of nature and the integrity and health of coastal communities and marine ecosystems. The main challenge for the environmental movement will be to develop, engage in, and transform the policies and politics of sustainability. Universities can play a critical role in capacity building and policy making related to this issue.

Form and Content of the Environmental Movement

The last three decades of the twentieth century saw the steady growth of the environmental movement in Puerto Rico—a movement comprising a wide spectrum of groups and organizations, with diverse memberships and political agendas.[4] I propose here a framework for understanding the environmental movement in relation to population, politics, policies, and the economy. Using a Weberian typology, we can classify the formal and informal environmental organizations working in the coastal zone into four general types. Conservation and environmental non-governmental organizations (NGOs) are composed of what Gouldner (1985) called the intelligentsia (scientists) and intellectuals (social scientists, artists, writers, and philosophers) who share a "culture of critical discourse" derived for the most part from their university experience. Members of the environmental movement at this level tend to subscribe to the New Environmental Paradigm (NEP)—a set of social values, attitudes, and practices that rejects the consumption-production paradigm of industrial society and an environmental ethic based in conservation values (Catton and Dunlap 1979). These NGOs receive funding from international and national agencies and foundations for environmental education and action and for conservation of natural areas. Environmental-social-political NGOs, in contrast, are formed by environmental, social, religious, labor, and political activists. Their political, religious, social, and environmental agendas and ideologies are complex and often intertwined. These NGOs provide technical and legal advice to communities, labor unions, and other groups that oppose either the state or the private sector in the environmental field. The third type, environmental groups, is made up of diverse community members (communities, clubs, NGOs) confronting a single environmental problem affecting the area. These groups are, at times, formed by a strategic coalition of some or all of the other groups mentioned above. The fourth type discussed in this chapter consists of community environmental organizations—local, grass-roots groups seeking to protect their own communities. These include organizations formed by resource users (e.g., fishers, boaters, farmers). Although these organizations are composed of specific sectors of society, they show a certain level of social diversity, depending on the demographics of their communities. These organizations may be either informal or formal. Usually they tend to formalize as they go through various steps in their development, increasing their tactical and strategic capabilities (after Hernández 1990).

What unites these types of groups into a movement is the fact that they are part of a process in which civil society challenges the state in an "environmental field" consisting of the processes and problems that affect the social, cultural, and biotic health of the community, alter ecosystems, threaten species, and change traditional culture and resource use patterns. This field is characterized by the exclusion of local communities and organizations from environmental policy and decision making. In response, the environmental movement engages in an ongoing process of empowering members of communities, groups, and institutions to confront the state. Environmental groups design and implement tactics and strategies, educate and organize communities, engage in co-management arrangements, claim traditional rights, and underscore the importance of local indigenous and traditional forms of knowledge and resource use (Acosta 1995; Renard and Valdes Pizzini 1994).

Like their counterparts throughout the world, members of the Puerto Rican environmental movement engage in legal, social, political, cultural, and economic disputes with industrialists, real estate developers, and the state when the latter threaten the health and integrity of social and natural communities. Contention in Puerto Rico takes place in different forms: campaigns for the conservation of protected and unprotected natural areas, "Not in My Back Yard" (NIMBY) opposition to projects, protests against "locally unwanted land uses" (LULUs) and disruption of traditional resource use patterns, and, most important, struggles to mitigate health and environmental problems caused by urbanization and industrialization. Recently, several organizations in Puerto Rico have engaged in co-management efforts that entail a sharing of responsibility and authority in the management and stewardship of natural areas and resources.

The development of the environmental movement in Puerto Rico parallels sharp growth in the island's per capita income, consumption, population, and industrial development and a dramatic collapse of the agriculture on the island, particularly in the coastal plains, once dominated by sugar cane and pastures. Urban settlements have encroached upon agricultural land, steadily expanding to produce "urban sprawl."

Changing Patterns of Land Use in the Coastal Zone

Urbanization patterns in the coastal zone, like those in industrial and post-industrial societies, feature dependence on automobiles and highway construction, commodification of space, construction of single-family housing units on the urban fringe, deurbanization of the central cities, and the hyper-mobility of capital (Gottdiener 1985; Knox 1993). Through the mid-1990s, urbanization and industrial development, population pressure, and the problem of waste disposal turned the coastal zone of Puerto Rico into an area of social and political contention. While environmental problems are present throughout the archipelago, including in mountainous inland areas, they are more acute in the coastal zone.

Population growth in the coastal zone of Puerto Rico began in the early twentieth century when large numbers of landless rural laborers abandoned the highland municipalities of Puerto Rico in search of work in the coastal zone. The U.S. invasion of Puerto Rico in 1898 and its aftermath, coupled with major hurricanes in 1899 and 1928, dramatically altered the island's power structure and weakened the agrarian economy of its interior. As a result of these changes, population grew in the coastal towns and cities that housed industrial sugar mills (*centrales azucareras*), most of which were owned by U.S. firms.

Government agencies, together with landholders large and small, persistently destroyed, drained, and filled coastal wetlands (Álvarez-Ruíz 1991; Giusti 1994) throughout the twentieth century in an effort to turn unproductive land (generally wetlands) into productive agricultural fields, to eliminate what were thought to be disease reservoirs (mangroves and swamps), to eradicate shantytowns and their physical environs (mangroves), and to turn idle land (wetlands in general) into space suitable for infrastructure development such as airports and harbors (Martínez 1994; Sepúlveda and Carbonell 1988).

The political and discursive foundations for these landscape transformations could be found in New Deal reconstruction programs and in the discourse of "Operation Bootstrap," the island's industrial development strategy, crafted in the 1940s. Puerto Rico was betrothed to the code of the old industrial paradigm, which offered the promise of a better future, albeit at the expense of environmental destruction. At the time, there was no environmental movement, no NEP, no hint of a pluralist movement to question what seemed to be unstoppable progress. Social movements and political actions were fueled by the controversial question of Puerto Rico's commonwealth status and controversy, by new forms of armed struggle (e.g., bombings of U.S.-owned businesses), by opposition to the Vietnam war, and by the university student movement, all of which were in consonance with social movements elsewhere.

Military Bases

Owing to geopolitical considerations, the coastal zone attracted U.S. military installations throughout the twentieth century (Rodriguez-Beruff 1988; Estades-Font 1988). Naval bases and installations were constructed in the island municipalities of Vieques and Culebra, in San Juan, and in the town of Ceiba (Roosevelt Roads Base), displacing many poor local inhabitants of those areas. Thousands of acres of mangrove forests and beaches were transferred to the U.S. military, and the use of some of these lands in Culebra (until the early 1970s) and Vieques for target practice became major political and environmental concerns (Delgado Cintrón 1989; Giusti 1999; McCaffrey and Baver, chapter 8 in this volume).

From 1976 through 1981, Vieques fishermen openly and consistently defied U.S. Navy prohibitions against entering target areas during bombing exercises. As a result of these highly visible acts of defiance, many political and resource-user organizations gained popular support throughout the island. In the early 1970s, Puerto Rican Independence Party (PIP) leaders, notably Rubén Berríos Martínez, joined the civilians in a confrontation over the U.S. Navy's occupation of portions of the island municipality of Culebra, and requested the cessation of its use as a bombing range. Culebra is no longer used for military purposes; base lands were distributed to the local government and to the U.S. Fish and Wildlife Service. Since the 1970s, other organizations have followed similar paths in confronting the U.S. military forces using political, peace, and environmental arguments. In the late 1990s, a massive pluralist protest tried thwarting installation of a Navy radar facility near Lajas on the southwest coast.

Similarly, Vieques fishers demanded a halt to military use of the area, arguing that military target practice was contributing to the environmental degradation of local coral reefs and coastal habitats. Some observers saw the fishers' protests and defiance of regulations as a unique instance of nonpolitical, nonpartisan confrontation with U.S. military authorities. In 1999, the death of a civilian guard due to a bombing accident triggered a massive protest and a formal request from the civilian community to end the use of Vieques as a bombing range and military post. A key argument supporting the request that the U.S. Navy leave Vieques was the claim that bombing caused contamination and environmental damage to the area. Bombing was indicted as the cause of unusual patterns of morbidity and disease among the local population. This well-documented (Giusti 1999; Benedetti 2000) process consolidated the interests and forces of the civilian community, including the environmental movement.

Agriculture

In the early twentieth century, the agricultural economy of the coastal zone was dominated by sugar and cattle grazing. However, despite increasing tobacco production by U.S. firms (Pumarada-O'Neill 1993), Puerto Rico's agricultural economy contracted steadily from the 1930s on, especially in the coastal zones. In the late 1980s, the amount of land devoted to agriculture increased slightly owing to government policies stimulating production of crops for local consumption (rice, tomatoes) and for export (millet, mangoes), but these efforts did little to reverse the steady decline of agriculture. While urbanization and industrial development brought new social and economic pressures to bear in conflicts over space and resources, agriculture played a major role in habitat loss and environmental deterioration in the coastal zone, processes that occurred quietly and imperceptibly without generating environmental opposition.

Industrial Development

Puerto Rico's economic and industrial "miracle" resulted from government investments in infrastructure and the success of tax-exemption mechanisms in luring stateside manufacturing firms and enterprises to the island to establish what came to be called "936 industries" (so named after the Internal Revenue

Service Code, Section 936). Industrial development occurred in several phases: garment firms were followed by electronic appliance manufacturers, food processing plants, heavy refineries, advanced electronics, and pharmaceutical plants (López Montañez and Meyn 1992; CIIES 1992). The latter phase of the industrial development process entailed production of large quantities of toxic wastes, which in turn led to acute health problems on the island, especially in the 936-Industrial Belt of the north coast.

The government controlled the supply of water and electricity, the latter generated by a network of petroleum combustion plants built throughout the coastal zone. In addition, dams and reservoirs were constructed for the water supply. The results of this process in terms of habitat destruction and introduction of exotic species have not been assessed. However, reservoirs came into the limelight during a recent "drought" when it was discovered that construction permits had been authorized for critical watersheds. Construction in those areas had caused erosion and the accumulation of sediment, which in turn had dramatically reduced the storage capacity of the reservoirs, thus contributing to the so-called drought.

The environmental movement has consistently criticized development in reservoir watersheds and has confronted both the state and the private sector on the detrimental impacts of infrastructure and industrial development on coastal areas. Citizens have also played a key role in environmental protection. For example, fishers and coastal settlers have opposed the construction of oil refineries on the south coast (Pérez 2000), as well as an attempt to build a coal-fired generating plant on the west coast (Maldonado 2000; Anazagasti 2000). The environmental movement against the coal plants in Rincón and Aguada in the 1970s and 1980s is typical of industrial societies: it was carried out by pure grassroots movements and community-based organizations, energized by community activists and other political actors. A recent study in the municipality of Guánica, on the south coast of Puerto Rico, shows how the local population formed a pluralist body made up of residents ranging from teachers to fishermen, and of all political tendencies, that has engaged in many struggles against industrial development and pollution (Acosta 1995).

Environmental Health Issues

Health and water quality are issues that have united many environmental organizations throughout the years. As noted scientist and environmentalist Neftalí García (1988) argues, one basic concern of the grassroots environmental movement has been the health of the local population, which has been impacted by industrial development and the ensuing contamination of the water and air. In the municipality of Cataño, south of old San Juan, a group of local residents organized a movement to protest the level of particulate matter emitted by an energy plant in the municipality managed by the Commonwealth's Autoridad de Energía Eléctrica (AEE). The Federal Environmental Protection Agency (EPA) designated the area as the Cataño Air Basin, and recognized the health problems caused by violations by the government energy company (AEE). Similar grass-roots organizations coalesced in the town of Guayama in the south and among industrial workers in the western city of Mayagüez.

Mayagüez also witnessed the rise of a grassroots movement called Mayagüezanos por la Salud y el Ambiente, which has been active in opposing the AEE's plans to build a coal-fueled electricity generating plant and to subcontract its operations to the Congentrix firm. Main issues in the campaign were the health of the nearby communities and the potential hazards posed by other projected plants. These and other efforts on behalf of health in the workplace and in the communities are perhaps the direct heirs of the fight to control contamination caused by oil refineries in Cataño in the early 1980s, in Guayanilla in the south coast in the 1970s, and by Union Carbide in Yabucoa in the east from 1973 to 1985.

Industrial contamination of freshwater and coastal bodies of water has been a major problem in Puerto Rico, affecting watershed areas, underground aquifers, and coastal waters. Environmental NGOs like Misión Industrial, consulting firms (Servicios Técnicos), state officials (Environmental Quality Board 1994), university programs such as Puerto Rico's Sea Grant College (Vélez-Arocho 1994; Chaparro 1998), international environmental organizations, and independent researchers (Hunter and Arbona 1995) have undertaken critical assessments of Puerto Rico's water problems, pointing to the role of illegal practices by industries and of waste disposal by communities and individuals. Local environmental groups like Mayagüezanos por la Salud y el Ambiente have taken the local tuna canneries to court to make them comply with the EPA's regulations related to waste disposal in the Mayagüez Bay.

Urban Sprawl

According to the U.S. Bureau of the Census, nearly 71 percent of the Puerto Rican population now lives in urban areas. During the first half of the twentieth century, urban expansion followed the archetypal growth pattern of the less developed countries: rural residents moved first to poor urban neighborhoods close to the city center, and later to marginal lands in the periphery. The formation of *arrabales*, or shantytowns, in wetlands on the urban fringe and on other idle lands became a familiar pattern (Ramírez 1976; Safa 1974). This process contributed to the depletion of mangrove forests, especially in the north coast (Martínez 1994).

The 1960s saw the growth of the construction sector, fueled by the massive expansion of subdivisions or *urbanizaciones* composed of mass-produced concrete and cement block houses—houses built for the islands to receive large numbers of return migrants during the 1970s. By the 1970s, government undertook urban renewal programs and the relocation of the urban poor into public housing projects called then *caseríos*, now *residenciales públicos*. Housing construction peaked in 1974 ($487.2 million). The construction sector stagnated from 1974 to 1977, and did not fully recuperate until the late 1980s (Villamil 1994,

14). According to official sources, investment in the construction sector reached $2.98 billion in 1994, a 5.6 percent increase from 1993. By 1995, public and private investment in construction totaled $3.3 billion, most of which went into construction of private housing, hotels, and shopping centers. A total of 13,664 housing units (mostly subdivisions) were constructed in 1995, and spending on housing construction grew 18.6 percent from 1994 to 1995 (Neggers 1995, B29).

This urban expansion, coupled with industrial development, was deemed responsible for the acute contamination of aquifers on the north coast. While available statistics are only approximate, but they show clearly that total population and the total number of manufacturers are positively correlated with well closings. Hunter and Arbona (1995) find a strong correlation between population, industries, and contamination of water wells in the limestone formation of the karstic zone of the north. The north coast is the main axis of growth for the San Juan–Caguas Metropolitan Area, a region that grew from a conurbation of nine municipalities and 1.2 million inhabitants in 1980 to encompass 30 municipalities and almost 2 million people in 1990. At this point, the metropolitan area extended from Barceloneta in the central north coast to Fajardo and Humacao in the east corner of the island. Suburban growth in San Juan has also reached the once rural and mountainous areas of a municipality like Caimito. The costs of this urban expansion, in terms of water quality and environmental pressures, are substantial. For example, the municipalities adjacent to the San Juan Bay estuary grew significantly from 1950 to 1990, adversely impacting that in-shore ecosystem. The poor communities of Caimito are now surrounded by middle- and upper-class subdivisions. Those projects have caused erosion and changes in the watershed that have had negative impacts on the San Juan Bay estuary. These older communities have repeatedly requested better environmental safeguards and a moratorium on housing development. Similarly, intensive urbanization in the Río Piedras highlands has caused degradation of water quality throughout the watershed (Gilbe 1998) and sedimentation of portions of the estuary (Webb and Gómez Gómez 1998, 5–6).

The environmental movement has been overwhelmed by the rapid pace of urban sprawl, and its actions have been primarily directed at the protection of the karstic environment. Another target of environmental activism has been the proliferation of resort hotel complexes and exclusive subdivisions that threaten the *poyales* (wetlands) located in Humacao in the east and in Dorado on the north coast (Alvarez Ruíz 1991). Other efforts, which have enjoyed broad media coverage and popular support, have sought protection of the Vacía Talega wetland in Carolina-Loíza, the coastal lagoon of Tortuguero in Vega Baja, the estuary of the Espíritu Santo River in Río Grande, and El Yunque (or the Caribbean National Forest) in Luquillo and various other municipalities, all of which are designated as protected areas.

Elsewhere on the island, both urban growth (although at a slower pace than in the San Juan Metropolitan Area) and development of infrastructure to support tourist-oriented development have drawn opposition from environmental groups and community-based organizations. One example is a government initiative to expand the Route 66 highway system in the northeast into environmentally sensitive areas.

Production of Tourism and Leisure Space

The sharp increase in the tourism infrastructure and recreational facilities in the coastal zone has sparked conflict between the environmental movement and the government and private sectors. The east and west coasts are the most affected by these projects (Valdés Pizzini, Chaparro, and Gutiérrez 1991). The Comité Pro-Rescate de Guánica, a pluralist community movement, was formed in 1986 to prevent construction of a Club Med facility on beachfront adjacent to the Guánica Dry Forest Biosphere Reserve. Despite the prospect of jobs and increased income for inhabitants of an area depressed by lack of industry, low levels of local economic investment, and the collapse of the Guánica sugar mill, the communities understood the importance of biodiversity conservation and recognized the need for environmentally appropriate development (Álvarez Ruíz and Valdés Pizzini 1990).

The Comité Pro-Rescate de Guánica campaign paralyzed the Club Med project, protected the dry forest, and encouraged the Puerto Rico Conservation Trust (a conservation NGO) to purchase the property for conservation. Despite these efforts, the Commonwealth recently rezoned the adjacent area as a tourism zone, making development on the borderline of the reserve a possibility. This change in zoning is being fought by local property owners, most of whom are from the upper class, take an open pro-environmental stance, and had joined in the movement opposing Club Med (Montes and Santana 1994).

Throughout the Caribbean, urban growth and tourism development proceed to the detriment of traditional fishing communities. In Puerto Rico, fishers are the group most directly affected by urbanization and tourist development in coastal zones. They have developed a critical outlook and have played leading roles in numerous environmental actions. In 1983, the National Oceanographic and Atmospheric Administration's (NOAA) Marine Sanctuary Program proposed the designation of a marine sanctuary on the southwest coast of Puerto Rico with conservation and recreational purposes. NOAA encountered fierce opposition from fishers and community organizations who felt that imposition of a federal marine sanctuary would curtail their freedom and penalize them. They argued that intervention was necessary to stop the principal agents of mangrove destruction and sewage disposal: absentee upper-class owners of second homes (*casetas*) that were illegally built along the shoreline (Krausse 1994).

The complexities of this particular case are discussed elsewhere (Valdés Pizzini 1990; Fiske 1992). Suffice it to say that the movement succeeded in halting the marine sanctuary. It also established the artisanal fishers as an important political force in negotiations and confrontations with Commonwealth agencies around protection of coastal waters. Ultimately, local fishers, with a genuine concern for the conservation of the stocks, joined local researchers in supporting the establishment of a Marine Fishery Reserve in La Parguera. Like their counterparts in La Parguera, fishers from Boquerón, San Juan, Fajardo, Vieques, and Salinas have formed organizations to protect local waters from the assault of leisure activities and infrastructure. Illegal use and occupation of the coastal zone by middle- and upper-class "squatters" has been a problem for the fishers in other areas of the south coast, including Guánica; Papayo in Lajas; Ponce; and Las Mareas in Salinas. Local residents, communities, and environmental organizations have taken a determined stand to stop the illegal use of the coastal zone and to protect access to the coast, often without success.

The Case of Río Grande

In the 1950s, Río Grande, a settlement near the northeastern coast of the island, was a decaying agricultural town in the midst of the northern sugar cane belt. In the piedmont surrounding the town, small landholdings survived on the scant lands that were not controlled by the U.S. Forest Service. Following the dramatic collapse of agriculture in the municipality of Río Grande, coastal lands were systematically transferred to the Commonwealth's Land Authority, which in turn ceded them to private developers who began to construct mega-resorts and high-income housing villas with golf courses for wealthy investors. The government removed coastal dwellers from what were considered "pestilent," unhealthy, and flood-prone mangrove areas. With the poor removed, two corporations bought most of the idle land in the coastal zone and undertook an impressive string of projects: large agricultural and speculative ventures, the Coco Beach subdivision, the Berwind Country Club, Río Mar Villas, Ríomar, Yunque Mar Resort, and private houses (Caraballo 1991, 13–18 and passim).

Since 1970, the municipality of Río Grande, on the northeast coast, has experienced a remarkable increase in population. A landscape of 14 kilometers of beaches in coastal plains and sugar cane plantations changed drastically owing to the voracious consumption of coastal lands and beaches by the resorts and housing developments. As a result, the Río Grande population lacked adequate access to their own local beaches, due to the absence of any public road that could provide physical or visual access to the coast. In fact, Río Grande citizens have no developed public beaches in the municipality. One of the few sites where local fishermen could moor their boats and land their catch was Las Picúas, a strip of idle agricultural land on the shore, but the company that owned the land sold it to a developer, who in turn parceled out the property in small plots. The plots were sold for individual homesites—mostly weekend and second homes—without the knowledge or consent of the Commonwealth Planning Board. Thus, access to the beach was curtailed and the fishermen lost one of the very few available strips of land along the shoreline.

The history of this conflict is complicated and sad, and the matter is still unresolved despite a court order that declared plot division and housing construction illegal and validated the fishers' right of access to the shore. The case of Las Picúas became a legal landmark, and it had a profound impact on the environmental movement. Consciousness of the problem of beach access and the ensuing confrontations among the state, the private sector, and the local population for the use and control of the coastal zone became evident in other cases throughout the island.

Las Picúas turned from a site and a case into a symbol of community resilience in the struggle of the coastal poor to regain access to the shore, access that tended to decrease with tourism and leisure development.

In Puerto Rico, access to the coast is at the core of what some call environmental justice. A number of environmental groups have claimed the shoreline as the last frontier for free leisure space and productive activities in close contact with nature. In summary, the Río Grande story illustrates the forces that resulted in the demise of agriculture, the rise of urbanization, and the growth of leisure and a tourism infrastructure, and the ways in which these processes have affected surrounding habitats and curtailed equitable access to and use of the coast. The rupture of "traditional rights" and the de facto violation of laws intended to impede the privatization and appropriation of beaches were key issues in coastal user conflicts at the turn of the decade, and they have continued to be so as environmental contentions escalate.

The Future of the Environmental Movement

It is difficult to assess the growth and development of the environmental movement in Puerto Rico due to a lack of sociological analyses. However, in this reflexive work, I offer a handful of thoughts on the movement's recent development and future path.

Environmental impacts on ecosystems and on human communities are of great magnitude and affect all orders of life. In a political context devoted almost exclusively to the status issue and the old paradigm of growth, the main moral responsibility for environmental stewardship has fallen on the shoulders of civil society. The Puerto Rican environmental movement was probably ignited by those communities affected by development projects and by changes in the overall health of the population resulting from acute industrial contamination.

Key actors in this process were the religious organizations and NGOs devoted to solving local health problems, improving the material conditions of poor communities, and empowering those communities, both politically and spiritually. These organizations provided communities with the tools for self-support and for political and civic organization and praxis, in their daily struggle for a better quality of life. Environmental organizations with long histories of projects for community empowerment and political struggles, such as Misión Industrial, tend to corroborate our assessment. The Puerto Rico National Ecumenical Movement (PRISA) also helped to forge what may be labeled as the Puerto Rican environmental movement (Saltalamacchia 1995).

The academic community and some leftist political organizations also participated in the formation of the environmental movement, and their contribution must be assessed. Leftist and pro-independence organizations and parties saw in environmental problems evidence of either colonial dependence on and exploitation by the metropolitan power, or the contradictions of capital. In my own assessment, their contribution to the environmental movement has been uneven; but a more comprehensive assessment must await further research. Recently, there have been discussions about their role in the struggle over those community problems and their participation as activists in a process characterized by pluralist, class, and political (ideological) coalitions (García 1988). The Pro-Independence Party (PIP) has been very selective in its environmental incursions. However, it has been involved in such highly publicized cases as the Vacía Talega struggles (Irizarry-Mora 1996).

Since the mid-1980s, several university programs and academic community members have increased their efforts to solve local environmental problems. This increase has been documented in the press (Rivera 1990, 1995), but a more critical assessment remains to be done. The academic "community" has intervened in a number of environmental actions over the past twenty years. In some instances it has played a key role in mobilizing protest and confronting the state with the technical knowledge and "culture of critical discourse." Certain successful environmental struggles in Puerto Rico may be linked to the insertion of a progressive episteme of academics in the environmental discourse. The university has also disseminated information and supplied the technical and social knowledge the community-based organizations need in their struggles. Finally, it has helped to consolidate regional networks of environmental and community organizations.

At present, the environmental movement features a large number of community-based organizations, environmental NGOs, and interest groups, and the university is an important participant. Although ideological discourses, social practices, and historical experiences tend to draw the different groups and organizations together, they remain scattered throughout the physical and political landscape. Most are site-oriented, focused on their particular problems, and unable to make the ecological and social connections between environmental impacts at one site or community, and their ramifications and linkages with the rest of the archipelago.

The pluralist backgrounds of these organizations and their individual and collective differences in cultural capital, class position, ideologies, political affiliations, and primordial loyalties keep them from forming a united front. As a result, the environmental movement remains fragmented despite the pervasive nature of environmental degradation. These constraints, combined with the status orientation of political discourse in Puerto Rico, have impeded the formation of environmental political groups similar to the Greens in Europe and in other parts of the post-industrial world.[5] Nevertheless, the formation of a coherent and united front against environmental degradation can happen.

Puerto Rico's Environmental Future

The environmental prognosis for the island is not encouraging. Recent trends suggest that Puerto Rican economic policy will continue to embrace the old industrial paradigm, with its bias toward growth and disinterest in the externalities produced by development. Sustainability is unlikely to be an important factor in planning and development. Patterns of urban growth, consumption, and use of resources and coastal land, coupled with timid government efforts to enforce environmental laws, low levels of fiscal support for environmental agencies, and half-hearted official efforts at conservation of critical habitats will force the environmental movement to continue with its stewardship of nature, and perhaps drive its diverse participants to coalesce into a united front.

In this section, I list ongoing processes that are likely to claim the attention and efforts of Puerto Rican civil society in the next decade. Depending upon how they threaten the integrity of coastal ecosystems and human communities, these processes will shape future environmental actions. When these summary remarks were first written in 2000, they reflected policies implemented by the New Progressive Party (NPP, a pro-statehood party), which had held power from 1992 to 2000. In 2001, a new administration, from the Popular Democratic Party (PDP, a pro-Commonwealth party), came into power. The development policies of the PDP government appear to have been quite similar to those of the NPP. The government of Sila Calderon sought to increase economic activity by promoting construction, tourism, and manufacturing. However, one critical difference that needs to be assessed and monitored is the PDP administration's commitment to sustainable development. The party's platform included an urban growth policy, a moratorium on construction in ecologically sensitive areas, development of a land use plan, installation of an ecologically oriented Planning Board, and other environmental policies related to waste disposal and citizen's participation. It is not clear, however, what impact these PDP initiatives had. This requires a thorough assessment from environmental organizations, policy analysts, and scholars.

Economic Trends

Puerto Rico is likely to remain committed to the "old industrial paradigm" for at least the next decade. The main lines of action identified in the Commonwealth's New Economic Development Model (NEDM), developed during the administration of Governor Pedro Rosselló, are tourism, industrial development, external trade, capital markets, human resources, and science and technology. We can also expect that in the future traditional uses of the coastal zone—agriculture, harbors, fishing, and industrial food processing—will be replaced by tourism and construction of leisure infrastructure, transshipment harbors, shopping malls, and technology-oriented enterprises. For example, Ponce lost its tuna processing plants in the mid–1990s. Similarly, Mayagüez is slowly losing its tuna cannery jobs. The garment industry, which attracted many rural workers to the west coast, is also disappearing (Griffith et al. 1995; Valdés Pizzini et al. 1996). Government reports indicate that a large transshipment harbor is planned for the Ponce, Guayanilla, and Peñuelas area, in southern Puerto Rico. As in the case of the Commonwealth Oil Refinery Company (CORCO) construction, changes in the local and regional economy, negative environmental impacts, and the transformation of traditional communities are expected to occur (Pérez 2000).

> *The Construction Sector.* In the last ten years, the Puerto Rican government has invested heavily in infrastructure projects to stimulate the construction sector. It has also instituted a policy of "fast tracking" procedures and permits to allow an increase in the number of projects receiving government approval. Stimulating construction was viewed as a way to offset a decline in the growth rate that occurred in fiscal years 2000 and 2001, following an "extraordinary boost" of 4.2 percent in 1999. Economic growth in 1998–99 was related to an unusual increase in construction projects due to reconstruction efforts after hurricane Georges. The extraordinary nature of this growth spurt notwithstanding, local economists see construction as the "motor for economic growth" on the island.

The present construction boom is expected to continue until 2006, when most of the government infrastructure projects will end. The economist Santos Negrón notes that in 2006 Congress is likely to eliminate all benefits enjoyed by U.S. industries under Section 936 of the Internal Revenue Service Code. According to Negrón, the Puerto Rican government established a strategy that calls for heavy investment in infrastructure until precisely that year, hoping to ameliorate the impact of a massive decline in employment in manufacturing. The multiplier effect of construction is greater than that for other economic activities because most construction materials are locally produced (cited in Díaz Román 1999, 10).

Housing, second home, and condominium construction on the shoreline is also expected to increase. Economic growth in the United States and Puerto Rico resulting in an increase in "disposable" income for the middle and upper classes is directly responsible for investment in apartments and houses for living and leisure purposes, as well as for investments to reduce taxes. The increasing number of condominiums and the high cost of those units is changing coastal communities. This increase is contributing to gentrification coupled with changes in traditional uses of the coast, displacement of settlers, and natural habitat destruction. Tourism, leisure, and recreational activities in the coastal areas are already taking their environmental and social toll. Any increase in those activities is likely to result in over-consumption of scarce water supplies, improper sewage disposal and associated pathogen contamination, solid waste problems as dumps reach capacity, and heavy seasonal traffic.

> *Tourism as a Priority.* As in Central America (Stonich 1999), tourism is a priority for Puerto Rico. The tourist development model set out in the NEDM and promoted by former Governor Rosselló (Lara Fontánez 1998, 4) and the

Commonwealth Tourism Company, emphasizes high quality of beaches (Chaparro 1998) and growth in the number of hotel rooms and tourist facilities.

Tourism was identified as the key sector with the potential to compensate for the loss of jobs in the "936 industries" and as one solution to the unemployment problem in the western region. The natural beauty of the southwest region, the quality of its beaches, its natural protected areas, and its recreational potential make decision makers prone to exploiting tourism as an economic alternative. Although a large number of foreign and U.S. visitors come to the region (mainly in winter), the majority of the visitors are local (Puerto Ricans) and flock to the region in summertime to go to the beaches and practice water sports. Unlike the tourism on the north and east coasts, which is dominated by large resorts, the industry in the southwest is still locally based, managed by small firms and local entrepreneurs. However, the Tourism Company has aggressively pursued a strategy to increase the number of hotel rooms in 2000 and 2001. Several new projects, in the planning stage or underway, are expected to produce more than 7,000 new hotel rooms (see Guadalupe Fajardo 2000, 20–26). At best, tourism is a highly voluble and uncertain activity,[6] and indicators suggest that tourism in the west coast is not sustainable in its present form and has a direct and negative impact on local resources. As is the case with house construction, additional tourism will create new pressures on the coastal zone, water supplies, and sewage and solid waste management. Access to the beach is likely to be limited, and beach erosion, littering, and pollution are all likely to increase as a result.

Persistence of the "Old Industrial Paradigm." These trends suggest that the government of Puerto Rico will remain committed to the "old industrial paradigm" for at least the next decade. To offset unemployment problems resulting from discontinuation of "Section 936" tax incentives for manufacturing, the Puerto Rico government, through the Puerto Rico Industrial Development Company (PRIDCO), is providing special tax credits for research, development, and training activities, and it is making concerted efforts to compensate for the acute loss of manufacturing jobs by attracting new firms and providing incentives to local entrepreneurs. This commitment to industrial development and construction is likely to worsen the environmental impacts documented by Hunter and Arbona (1995), unless a connection is made between new technologies and resource conservation. A challenge for future administrations will be to make economic growth compatible with sustainability.

Land Use Changes

The Fate of Public Lands. Privatization was an essential component of the NEDM. Privatization of government-owned lands is a small, inconspicuous component of that policy. During the Rosselló administration, the government transferred and sold Commonwealth lands to private developers, often at prices below market value. In one case, a plot sold by the Commonwealth lay within a mile of the Tortuguero wetland, a natural reserve under the protection of the Department of Natural Resources. If not monitored by civil society, tourism development could easily threaten the integrity of ecosystems.

The number of acres of public land accorded "natural protected area" status was, ironically, highest during the 1990s. However, most of this area was inland. No protected areas received sufficient funding for management or even for the development of management plans. Thus, management and enforcement of the Commonwealth's protected areas remain critical problems. Lack of funding may impede future work to establish marine reserves, to craft appropriate legislation for protection of coastal and offshore habitat (e.g., coral reefs) and fisheries (essential fish habitats), and to enforce Coastal Zone Program regulations.

Urban Growth. The number of people living in urban and coastal areas is expected to increase. As indicated by the 2000 U.S. Census, the San Juan–Caguas MCSA is expected to see growth in population and housing density, and a small but significant increase in the number of municipalities included in the area. This growth is expected to affect the surrounding wetlands and watersheds.

Civil Society Involvement in Environmental Protection

With Puerto Rico's main political parties locked in a struggle to resolve the status question, and with the government committed to promoting economic growth, stewardship of natural resources and protection of natural areas, watersheds, and biodiversity will increasingly fall to civil society, mainly those environmental organizations with proposals for co-management and management of "natural" spaces. This will require the empowerment of communities and organizations, and a knowledge of resource management technologies. In this section, I identify alternative paths that the environmental movement could traverse to have an impact on conservation of coastal and marine ecosystems and resources.

According to Hunter and Arbona (1995), the pristine nature of the island has been lost forever, due to the lack of implementation of conservation measures and policies. They cogently argue that "what is fundamentally needed is a wider level of public awareness and a consensus to act in harnessing the political will" (1995, 444). Their assessment underscores the desperate need for the environmental movement to have a stronger voice, to deploy a coordinated political effort, and to engage in strategies and practices that prioritize conservation and sustainability in a holistic manner. In conclusion, I argue that the following issues, processes, and potential activities need to be addressed by the environmental movement: population and consumption, information and technology transfer, policy making, sustainable development, a new economy, and traditional coastal communities.

Population, Consumption, and the Environment. A critical appraisal of urban growth and the construction sector is sorely needed. The relationship between population, consumption, and the environment needs to be assessed, employing critical theoretical perspectives that allow investigators and organizations to make the connections among those processes and variables. In Puerto Rico, the environment, as a "natural" category belonging to the physical world, is an object of analysis and research. As a space, it is also the object and field of political actions. However, little attention is paid to the triangular relationship among the environment, human populations, and consumption patterns, all of which are affected by economic policies. This triad is the focus of a growing body of interdisciplinary political ecology and applied research aimed at reshaping policies that affect biodiversity, ecosystems, and human societies (see Arizpe and Velázquez 1994; Stedman-Edwards 1998; Stonich 1998). In Puerto Rico, this triad must be assessed with special attention to the process of urbanization.

Information and Technology Transfer. The university has a role in empowering coastal communities affected by these processes; it can provide communities and stakeholders with the appropriate knowledge, technology, and tools to achieve their goals. Academia must also play a key role in promoting sustainability, environmental protection, and the conservation of biodiversity, ecosystems, and those human communities whose histories have been closely interwoven with the environment. Debates pitting the ivory tower against advocacy and objectivity against social action can constrain and modify university interactions with the public. But without informed and active advocacy, change will not occur. An example of this is the UPR Sea Grant College Program's project with local community organizations facing environmental problems. Ana Navarro, a Sea Grant water quality specialist, has developed a series of activities and interventions aimed at information transfer, capacity building, and empowerment of communities in western Puerto Rico (Navarro and Navarro 2001). The success of this project is likely to trigger similar projects for other coastal communities facing similar environmental challenges.

Policy Making. Information transfer must also continue at the level of policy making. Some government officials still believe that the university, like the community, should not have a role in policy making. This belief reflects a distorted view of democracy, a view the environmental movement has tried to dispel. Civil society and the university have the right and the duty to contribute to the welfare of the polity by offering appropriate alternatives based on the best knowledge and experience available. In informing various stakeholders, fellow citizens, and policy makers of the best alternatives and courses of actions, members of the university community must at times move away from the mainstream extension philosophy to play an advocacy role. For example, Puerto Rico lacked a firm policy on beach conservation and management until the University of Puerto Rico Sea Grant College Program published Chaparro's (1998) policy paper. Despite opposition and even political pressure from the Secretary of Natural and Environmental Resources, Chaparro continued to inform the public, stakeholders, and law makers of the importance of beach management and the full range of available options and programs. The results of his efforts were new legislation, formation of a board responsible for beach management, community and private-sector support for development of beach management projects, and the full (and verbatim) incorporation of Chaparro's guidelines in the DNER document on coastal policy, *Puerto Rico and the Sea* (Department of Natural and Environmental Resources 1999).

Sustainable Development. It is telling that sustainable development is absent from policy and from daily political and environmental discourse at all levels of Puerto Rican society. Elsewhere in the Caribbean, the discourse of sustainable development, protection of biodiversity, protection of ecosystems, and appropriate development for small islands is common. Ironically, the United States, our key political example and beacon, has a presidential policy on sustainability (unchanged under the Republican administration) that theoretically translates into action in the educational and resource management fields. For example, the National Oceanic and Atmospheric Administration (NOAA) and related programs like Sea Grant and the National Marine Fisheries Service feature a strong commitment to sustainability. Sustainable development is not a panacea for all environment-related ills, nor does it imply a well-defined set of rules that can be followed blindly. On the contrary, it remains a highly controversial and complex proposition that requires analysis, research, and experimentation through pilot programs. Nonetheless, focusing on sustainability forces society to deal holistically with a number of environmental issues, including waste reduction, conservation of nature, biodiversity and ecosystems, the wise use of resources, optimization of industrial and agricultural production systems, and improvement of commercial practices. It implies inclusion of communities in the conservation process, thorough planning and allocation of space and scarce resources, and a philosophical commitment to protecting the integrity of nature for future generations. These challenges must be met by the society and government of Puerto Rico.

The "New Economy." The Commonwealth's economic policy from 1992 to 2000, based on the NEDM, was an essential component of the New Progressive Party platform, and a key element of the social and economic policies designed to achieve statehood (Hexner and Jenkins 1998). In assessing the future of the economy, the NEDM emphasizes the role of science and technology, the promise of "the new economy" based on globalization, and digitization of production, commerce, and services. Interestingly, the political platform of the PDP contains an economic development strategy that is similarly based on the tenets of "the new economy." In my view, "the new economy" model does have the potential to promote resource conservation and sustainability, although its proponents have not noted this connection (Atkinson and Court 1998). Given the lessons of top-down economic development based on the "old industrial paradigm," commitment to the new economy could provide a new field of action for the Puerto Rican environmental movement that can, and ought to, seize the opportunity to promote alternative forms of growth that are consonant with ecosystemic integrity. In theory, the "new economy" should promote optimization of production processes, reductions of scale, new forms of transportation, use of telecommunications, digital transactions, and waste reduction, all of which are fully compatible with sustainable development principles. Once again, responsibility for making intellectual and concrete connections between "the new economy" and environmental protection will lie with civil society.

Traditional Coastal Communities. Local communities have an important role to play in the sustainable development of Puerto Rico's coastal areas. They must, and should, have a future in the context of current economic policies and the touted benefits of "the new economy." I argue, following Mowforth and Mount (1998), that sustainable tourism is consonant with the principles of the NEDM. Small, traditional, culturally bound communities are not antithetical to a globalized information economy dominated by post-Fordist practices. Despite attempts to homogenize local cultures (García Canclini 1999), these communities remain potential spaces for sustainable development and cultural integrity. Writing on communities of Chesapeake Bay and Albemarle and Pamlico Sounds in North Carolina, Griffith (1999) argues that fishermen and coastal settlers are the custodians of a long history of reciprocal exchanges with nature; that these exchanges are based on a set of rules very different from those that guide markets; and that their behaviors are closer to what some call sustainability. These populations also possess a rich knowledge of nature, which is critical to its understanding, appreciation, and conservation.

From La Parguera to Vieques, Puerto Rican fishers and traditional coastal settlers remain the fundamental custodians of the coastal environment. Ironically, in playing this historical and cultural role, they are perceived by policy makers and others as barriers to progress. Puerto Rican fishers and coastal settlers are also stewards of a vast body of knowledge related to habitats, species, and ecological dynamics, as well as of a historical perspective on changes in nature (Valdés Pizzini et al. 1994; Giusti 1994). Sea Grant–sponsored research indicates that fishers' patterns of territoriality and spatial utilization are based in old conservation practices and ethics and that their behaviors are rooted in an idea of the commons entailing active community participation in defining sea tenure and usage patterns (Jean-Baptiste 1999). More important, fishers have historically sought to resolve coastal conflicts in ways that favor environmental protection, and I believe fishers and coastal communities will continue to play this role, defying economic trends as well as experts' predictions of their demise.

Final Comments

The coastal zone of Puerto Rico has been, and will continue to be, a critical area for environmental contention between civil society, the private sector, and the state. This chapter is a first step toward understanding the social, political, and economic processes at work in the coastal zone. As civil society organizations prepare for the future, they will need a well-documented history of the environmental movement and its relation to the island's socioeconomic processes. Also required are a more thorough critical assessment of the environmental situation and a presentation of alternative scenarios, much like those concerning the status issue and the economic future of Puerto Rico (see Gutiérrez 2000). So far, the environment and its social, political, and economic ramifications have seldom attracted social scientists or interdisciplinary teams. I certainly hope this situation changes, as the future could be shaped by the strategies such analysis could provide.

Notes

An earlier version of this article titled "Historical Contentions and Future Trends in the Coastal Zone: The Environmental Movement in Puerto Rico," was produced by the University of Puerto Rico Sea Grant College Program at Mayagüez (Publication No. UPRSGCP-R–80).

1. This chapter is dedicated to the memory of Carlos Goenaga, Pedro Santana Ronda, and Carmen Salomé Rodríguez, to whom we are deeply in debt. For their critical comments and suggestions, I thank Ruperto Chaparro, María Benedetti, and Miguel Sarriera. I also benefited immensely from conversations on this topic with Migdalia Álvarez Ruíz. I am deeply indebted to Jaime Gutiérrez Sánchez for collaboration in this research project and for coordination of field activities. This work was also made possible by the fieldwork and commitment of Marialba Hernández and María del Carmen Caraballo. Editors María Benedetti and Andrea Torres did an excellent job in making my ideas intelligible.

2. Since 1992, large number of students have participated in a variety of investigative endeavors on the following topics: municipal planning, local community-based environmental organizations, coastal conflicts reported in the press, changes in the demographics of coastal districts (*barrios*) and municipalities, inventories of businesses, and a census of business owners' opinions on a variety of coastal issues affecting their livelihood, the environment, and the local economy. This last teamwork effort also produced several case studies of communities, using ethnographic and survey data to assess social displacement and changes in the social composition of coastal settlements. Case studies from other municipalities, such as the northern coastal town of Río Grande, were also prepared for this project. In the municipality of Guánica, we had the opportunity to link our project to similar efforts and activities conducted by other colleagues and students (see Álvarez Ruíz and Valdés Pizzini 1994; Montes and Santana 1994; Acosta 1995).

    This chapter is based on data collected during this project by student and faculty researchers on the role of environmental organizations in the west coast of Puerto Rico. The original piece of research was conducted by Marialba Hernandez, whose work serves as a platform for this article (Hernandez 1990). I have expanded the scope of that initial research to provide a sociological assessment of environmental conflicts in the coastal zone in Puerto Rico as a whole.

    Originally, the objective of this report was to provide a simple typology of organizations and a brief description of their activities. I later added sections on socioeconomic trends in Puerto Rico and on the alternative paths down which the environmental movement may go in the next ten years, in relation to population, politics, policies, and the economy.

3. For an exhaustive inventory of events, groups, and organizations forming the Puerto Rican environmental movement, see Giusti (2001).

4. Exceptions are two general works (Torrecilla 1986; García 1988), several specific case studies (Acosta 1995; Anazagasti 2000; Maldonado 2000), and an overdone critique (Cerame-Vivas 1994).

5. Lack of block support for an independent green candidate for the Senate (the environmentalist and scientist Neftalí García) is perhaps an example of this.

6. For example, in the summer of 2000, local firms complained about the scarcity of tourists in the region and the lack of business.

CHAPTER 5

The Struggle for Sustainable Tourism in Martinique

MAURICE BURAC

TRANSLATED BY JULIET MACDOWELL AND ROBERT C. A. SORENSEN

*T*he development of tourism, and the use of land to support it, increasingly preoccupies public opinion in Martinique. The decline in agriculture, the rise of the service sector, and the willingness of the French state to follow a policy of economic diversification have all contributed to a boom in tourism and cruise-ship activities. The demand from private investors for the best coastal sites, environmental damage resulting from tourist uses, and the numerous contradictions in government policy have produced a population with a heightened sensitivity to environmental matters and have stimulated the growth of a number of ecological organizations.

In the 1970s, 1980s, and 1990s, various movements were organized and various actions taken by ecological groups to convince economic elites and public officials to move from a traditional concept of tourism to a more modern view of sustainable tourism. In the process of development of a regional territorial management plan for Martinique between 1993 and 1995, environmental organizations offered several proposals to stop or modify tourism development programs proposed by private investors and supported by elected local politicians, by the French state, or both. This chapter examines both past and more recent efforts to develop an environmentally sensitive tourism sector in Martinique.

Tourism Development: Data from Martinique

From Agriculture to Tourism

Since the 1970s, economic changes have accelerated on the island. Agriculture, once the primary sector, has declined. Like other Caribbean islands, Martinique has experienced an especially marked decline in sugar cane production. Even though the rum and banana industries have survived (albeit with serious difficulties), the general tendency points toward disengagement from agricultural activity. Rather quickly, the tertiary or service sector has become the most dynamic in the economy. State policies and private initiatives favoring service sector development have produced the current situation, in which services represent almost three-fourths of Martinique's Gross Domestic Product (GDP) and furnish the bulk of the island's employment.

In the 1970s, following the examples of other Caribbean islands, officials emphasized tourism activities in various social and economic development plans. The French national government and local public entities increased investments in infrastructure, including roads, airport modernization, and installation of running water, electricity, and sewage systems throughout Martinique's villages. In addition, private companies investing in the tourist sector benefited from various customs and tax incentives. In 1986, a very favorable incentives package was adopted to encourage even more private investment in tourism. Public officials, pushed by conservative political parties, sought a transformation of the economy from plantation agriculture to a diversified economy in which industry and, more importantly, services would generate wealth and employment. Tourism was at the center of this plan.

The financial and political push from the state, the interest of the private sector, the decrease in airline prices between France, the rest of Europe, and the French Antilles, and the desire to compensate for the decline of the North American tourist clientele (from both the United States and Canada) with an increase in European clientele were all factors that coalesced to assure a boom in tourism. The number of hotels and guest houses on the island grew rapidly. Nowadays, a varied and plentiful group of mainly European (especially French) tourists visit Martinique each year.

Public Mistrust of Traditional Tourism

The policies of the French state, which involved a retreat from the traditional sugar industry partly achieved by a decrease in subsidies to sugar enterprises, were felt until the 1980s. All private sugar plantations were closed, and the one remaining unit was transformed into a mixed enterprise, with public-sector participation in its financing and management. The national political parties, autonomists and independents, and the unions refused proposals made by the state while at the same time denouncing the private sector's poor management of the sugar industry. These opposition forces disapproved of the government's new development strategy, which they believed would create an even more artificial and externally dependent economy. Local sugar producers also tried, in vain, to fight against the loss of sugar subsidies.

In 1974, the presentation of an ambitious tourism project involving 11,000 beds in Sainte Anne, in the south of the island, provided the occasion for the local opposition to express its distrust of the state's new tourism policy and of public officials more generally. This project allowed private investors who had signed favorable deals with local Martiniquan politicians and representatives of the French state, to take over an area on the coast that was both attractive and popular among the locals. Furthermore, the site chosen for this new project, Les Salines, was not very far from another premier coastal area, Le Pointe Marin, where the French group Trigano had already constructed a Club Med vacation complex.

A campaign was begun by a group that called itself the Committee for the Defense of l'Étang des Salines and was made up of representatives from all sectors of the opposition. In part because of the large base of support, the effort was a complete success. The investors abandoned the project, to the great relief of much of the population, even though they realized that this environmental victory meant the loss of an opportunity to create thousands of jobs.

Martinique faced a difficult economic, political, and social situation in 1974. That year was marked by serious strikes in the agricultural sector, to which the government responded with violent repression of the workers' movement. This context of economic decline and political repression actually enhanced the stature of the Étang des Salines Defense Committee. The government, fearing even more disorder, began to pull back its support for the project, and ultimately the investors withdrew the entire proposal.

Several elements explain the widespread mobilization against the Salines project. Most important was the exceptional organizational effort by the leaders of the Defense Committee, a group composed of civil servants, professionals, and citizen representatives. Simply put, all were opposed to the use of the site. Also, at that moment, politicians had a real fear of more instability and of a local backlash against the growing foreign-born, European population that was settling on the island. Soon after his election French President Valéry Giscard d'Estaing announced his intent to implement a policy of economic departmentalization to reduce Martinique's social instability. Giscard called for an aggressive tourism development program that also implied an increase in immigration from France. However, the idea of transforming Martinique into a tropical tourist destination for the European French and facilitating their immigration to the island was perceived by many natives in an extremely unfavorable light, since many Martiniquans were already considering emigrating themselves because of the territory's high population density.

The failure of the Salines project ended the public's unquestioning acceptance of tourism development as well as the private sector's unquestioning willingness to invest on a large scale in such ventures. In fact, it encouraged the anti-tourism front to remain vigilant and relatively united until the 1980s. At that time, the Autonomists gained support from French President François Mitterand's Socialist Party and, somewhat ironically, because of their new local responsibilities, they began to play the tourism card as a factor in their development calculus.

Public Condemnation of Excessive Tourist Expansion
Several factors explain public mistrust of traditional tourism. With its 420 square miles, Martinique is heavily populated, and its ecosystems are fragile.

The island faces an ever-present risk of natural disasters (e.g., volcanic eruptions, earthquakes, hurricanes, tropical storms, and floods) as well as man-made damage. In addition, Martiniquans have lacked a well-conceived development plan and have as a result experienced (1) chaotic urbanization and its accompanying dysfunctions, and (2) political and cultural assimilation. Martiniquans often heard conservative politicians speaking of Martinique as France—as part of the metropole. It is not surprising, then, that the inherent economic and social contradictions of colonialism regularly play themselves out as popular dissatisfaction or even rebellion in periods of crisis.

The shortage of housing and the demand for land have created a strong squatter movement and squatter settlements dot public lands on the coasts. Another factor explaining public rejection of large tourism projects is of the public memory of beach use conflicts in the early years of tourism development. Several conflicts occurred between local residents and hotel management when the first hotels, typically at prime locations on the coast, closed off access to certain beaches by erecting barbed-wire fences or simply prohibiting non-guest use of the beach.

The lack of reflection on these conflicts by public officials contributed to public distrust of the tourism sector. In short, during the 1970s and 1980s, tourism seemed like a new form of human pollution within a structure of neglect. The fragile community of Martinique, which depended on yet at the same time was vulnerable to the outside world, was imbued with a strong spirit of rebellion, pushed first by the Autonomists and later by the independence supporters who feared that the island would lose its soul once it was given over to foreigners and tourists. Increasingly, these political groups pushed for action by ecological associations; the associations became much more combative in the 1980s.

The Struggle for Sustainable Tourism

The Affirmation of the Environmental Movement

At the end of the 1960s in France, officials responsible for planning and urban development placed special emphasis on local development. They stressed the need to make communities or groups of communities work together to design a framework to enhance their standard of living. The 1967 *Loi d'Orientation Foncière* (LOF) created new tools for controlling urban development. This, along with the *Plans d'Occupation des Sols* (POS) and the *Schémas Directeurs d'Aménagement* (SDA), would force both elected officials and the French public to consider the advantages of protecting nature and to question the environmental impacts of projects.

Full application of the LOF in Martinique in 1972 quickly sensitized public opinion to the constraints that islands pose. Applying a regulation of a developed country, where the laws are generally respected, to an underdeveloped overseas department where, for all sorts of reasons, laws are often not implemented, immediately exposed the limits of state intervention.

New French environmental laws enacted in 1976 paid particular attention to coastal zone management. Their delayed implementation in Martinique and the lack of public discussion, coupled with the adoption of ecological concerns by several popular organizations as part of their political agendas meant that protection of the environment in the territory would be more likely to come from ecological associations and political radicals than from specialized government institutions.

Between 1974 and 1976, mainstream institutions such as the *Société pour l'Étude, la Protection et l'Aménagement de la Nature a la Martinique* (SEPA-MAR), created as part of the larger *Société pour l'Étude, la Protection et l'Aménagement de la Nature dans les Regions Intertropicales* (SEPANRIT) *de Bordeau-Talence*, or the *Association pour la Protection de la Nature et de l'Environnement* (APNE), joined together to form the *Association pour la Sauvegarde du Patrimoine Martiniquais* (ASSAUPAMAR), *Comité de Résistance à la Destruction de la Martinique* (CORDEM), and *Association de Défense du Patrimoine Martiniquais et des Mal Logés* (APPELS). In the 1980s, these groups represented more structured movements in which identity conflict and environmental mobilization went hand in hand. These organizations gained so much credibility that the state gave them the right to intervene in the name of environmental protection, especially in legal questions. An agreement granting this authority was signed with APNE in 1979, with ASSAUPAMAR and SEPAMAR in 1986, and with CORDEM in 1988. The *Société des Galeries de Géologie et de Botanique, Découverte et Protection de la Nature* of Forte-de-France and the *Asociation Martiniquaise d'Initiation a l'Environnement* (AMIE) signed agreements in 1990 and 1993 respectively.

At present, the environmental movement consists of two types of organizations: (1) associative structures that are either autonomous or attached to a public institution and are moderately militant; and (2) a group of three associations that use more confrontational tactics. This group includes ASSAUPAMAR, CORDEM, and APPELS. The first group brings together more than a dozen associations, including SEPANMAR, APNE, the *Société des Galeries de Géologie et de Botanique*, and AMIE, all of which have benefited from agreements with the state. Also in this first type are other groups that vary in their levels of activism, such as Tabulikani, *Union Régionale pour la Gestion des Espaces Naturels et la Connaissance de l'Environnement* (URGENCE), *Fédération des Associations de Protection de la Nature* (URAPEM), *Amis du Parc Naturel Régional, Association pour la Promotion de l'Architecture*, etc. The main objective of these associations is to inform and educate students and the general public. Their involvement in different types of activities on behalf of the environment makes a general contribution in sensitizing public opinion to the benefits of nature and to considering these benefits in our daily lives.

Among the more activist organizations, ASSAUPAMAR has been for the last decade and continues to be the most influential force in the ecological struggle, especially in its efforts to promote sustainable tourism. Made up of a core of well-organized militants, including former active opponents of the 1974 Salines project, ASSAUPAMAR has paid special attention to land use issues. If, at the beginning, the organization used all means to oppose development projects seen as compromising the harmonious development of the island, it has had to modify its strategy in the 1990s. In trying to be more effective in the area of ecological protection, the group has sought to learn more about environmental law and the functioning of the French legal system. It has progressively abandoned land occupations, a strategy since adopted by the political movement MODEMAS. Furthermore, in instances when ASSAUPAMAR has opposed local or state governments or private investors, administrative tribunals have found the cases they presented to be well-crafted and persuasive. ASSAUPAMAR has successfully sued municipal governments for disregard of French environmental law or the urban development code.

We suggest that the frequency of legal battles related to urban growth on the island is linked to poor implementation or doubtful, or even irresponsible interpretation of national laws and policies regarding environmental protection and urban development. By obtaining injunctions from administrative tribunals, ASSAUPAMAR has blocked openings of shipyards and docks on the coasts, projects that were judged to have been improperly authorized. In particular, ASSAUPAMAR has also been active in denouncing the policy of total tourism, coastal degradation by private investors, subversion of regulations involving coastal construction projects, complicity in these acts by local elected officials, and the state's general irresponsibility. In sum, ASSAUPAMAR has played a prominent role in stimulating debate on the general issues of land management and sustainable development.

A second group, CORDEM, is closely linked to the Martiniquan Independence movement (MIM), which at times has controlled close to 22 percent of the seats in the Regional Council. CORDEM has less support than ASSAUPAMAR in the territory.[1] The organization has taken quite radical positions against state-supported private investment in the tourism sector and has used radio programs to make its positions known to the public. CORDEM regularly emphasizes the need to inventory Martinique's natural resources as well as its economic, cultural, and historical patrimony and to promote sustainable development.

Finally, born in a split with ASSAUPAMAR in the early 1990s, APPELS has also contributed to sensitizing public opinion to the dangers of unchecked or total tourism. Its information campaigns have stressed the dangers of anarchic urban growth, the problem of substandard housing, the lack of clean drinking water, pollution, risks associated with natural disasters, especially earthquakes, and new technologies.

The public's awareness of environmental questions has been reinforced because the relatively high standard of living enjoyed by a good number of Martiniquans has allowed them to travel in the Caribbean, to North America, and to Europe. They are therefore in a position to appreciate both the environmental problems and the conservation accomplishments in other countries. Scientific reports produced by experts, contributions by academics, information collected by state agencies and local groups such as *Association des Professeurs de Biologie ou de Géologie* (APBG) or *Société des Galeries de Géologie et de Botanique,* provide increasing amounts of material that sensitizes the public to Martinique's environmental realities. Under these circumstances one can understand the public's expectation that the island will develop a policy of sustainable tourism.

The Debate over the Regional Management Plan

In 1995 ASSAUPAMAR, CORDEM, and APPELS made significant contributions to the wording of the *Schéma d'Aménagement Régional* (SAR), a planning document intended to establish a basic framework for development that included evaluation of the land-use and environmental impacts of projects. These groups expressed profound reservations about the plan adopted by the Regional Council and made counterproposals geared toward development that would be more balanced and more appropriate for the Martiniquan context. In particular, the groups criticized the absence of an over-arching development plan, neglect of the recommendations of the 1992 United Nations Conference on Environment and Development (UNCED) on sustainable development, disregard of the urban development code, and an absence of strategic planning. These activists called upon elected members of the Regional Council to modify the plan dramatically and even urged the Council to rule on the plan's legality. The groups argued that the proposals concerning nature tourism compromised the idea of sustainability—bequeathing to future generations a healthy environment that promotes a high level of social and economic well-being.

Because of the large-scale use of natural spaces and the potential for irreversible destruction of nature inherent in most tourism projects, and also because Martinique's tourism sector is controlled largely by foreign interests, the ecology movement is quite skeptical of this economic activity. It was argued that the tourism option as conceived of in the SAR could not be considered a productive sector, because its prosperity would depend on a complex set of parameters and because it would offer limited possibilities for stimulating the Martiniquan economy. ASSAUPAMAR, in particular, has argued that the increase in hotel capacity as conceived of in the *Schéma* is so great that it would create intolerable pressure on Martinique's coastal areas, damage the island's scenic beauty, and cause pollution linked directly to the growth of mass tourism. ASSAUPAMAR also reaffirms the need to promote popular participation in all kinds of decisions relating to the island's future.

In its rhetoric CORDEM stresses that land use decisions should be made under the best possible conditions; this would require undertaking a comprehensive inventory of Martinique's resources and potential, as well as a precise statement of its constraints (e.g., major natural disasters), and that such an inventory should be used in all future development planning. Proposals have been offered by the groups for a multi-sector development program. A number of basic critiques from CORDEM and APPELS have reinforced the ASSAUPAMAR view that the SAR lacks a global vision of planning, and that, in general, it is ill-conceived and does not serve the long-term interests of the country. CORDEM has argued that the number of existing hotels and lodgings is largely sufficient for the coming years, especially given the current difficulties of certain large hotel chains. So CORDEM, along with the group of patriots from the Regional Council, demanded suspension of the new tourism projects in the SAR, including those against which ASSAUPAMAR had already begun litigation regarding the locations chosen. CORDEM further argued that freezing large-scale tourism operations would enhance the profitability of existing facilities. APPELS, too, joined the public combat and emphasized the need for designing a coherent program to promote sustainable development, one that would take into account the island's specific economic, social, cultural, and environmental conditions.

Conclusion

Martinique is an island where, for economic, political, social, and cultural reasons, the ecology movement has benefited from the vacuum created by the weak functioning of the institutions formally charged with protecting the environment. The diversity among organizations has allowed the population to benefit from ongoing data collection efforts, to share experiences regarding land use issues, and to develop a critical approach and a willingness to question the type of tourism that has been encouraged. Despite some misjudged interventions and some less-than-rigorous analyses, the movement has forced the state and the private sector to abandon projects scaled inappropriately for the island's size and density, and to modify their proposals in favor of greater sustainability. The battle is far from over for these organizations, especially as underdevelopment and chronic unemployment persist. Nevertheless, thanks to the legal arguments of the ecologists and to an increasingly attentive citizenry, public institutions have been progressively forced to apply the law—a fundamental step in the advancement of sustainable tourism.

Note

1. The Regional Council comprises Martinique's local Legislative Council plus the locally elected Deputies and Senators who represent the territory at the national level in Paris.

PART III

Behind the Beach

PRODUCTIVE LANDSCAPES AND ENVIRONMENTAL CHANGE

CHAPTER 6

Puerto Rico

ECONOMIC AND ENVIRONMENTAL OVERVIEW

NEFTALÍ GARCÍA-MARTÍNEZ, TANIA GARCÍA-RAMOS,
AND ANA RIVERA-RIVERA

*H*umans modify nature more than any other species. Their interaction with nature takes place within a social milieu that comprises scientific, technological, economic, political, ideological, and living and non-living nature-derived elements. The Puerto Rican economy and the social and natural components of its environment have been drastically transformed during the past century. In this chapter we offer an overview of these changes.

Puerto Rico is a subtropical archipelago that includes a main island and the smaller islands of Vieques, Culebra, and Mona. Its total area is about 3,435 square miles. It is the easternmost island of the Greater Antilles. Military strategists have coveted its geographical location since the Spanish arrived in 1493.

From Agriculture to Industry

At the time of the U.S. military invasion in 1898, Puerto Rico, until then under Spanish rule, was in transition from a pre-capitalist to a capitalist socioeconomic formation. Before 1898, the United States was the primary market for Puerto Rican products, followed by Spain. Coffee and sugar cane production were the main economic activities; subsistence farming and artisan fishing were also important. A banking sector had begun to emerge a few years earlier.

At this time the U.S. socioeconomic formation was characterized by increasing monopoly control of capital investment in oil, sugar, coal, and railroad transportation, among other economic sectors. Some of the first economic policies pursued by the U.S. government were currency change from the Spanish peso to the dollar, elimination of trade barriers between Puerto Rico and the United States, and suspension of credit. The Puerto Rican government's autonomous relationship with Spain was abolished, and the archipelago became a U.S. possession, a relationship that prevails up to the present.

Puerto Rico was thrust directly into the North American economy without any protection. It was only a matter of time before this would precipitate the destruction of pre-capitalist economic activities, particularly subsistence farming, and the consequent social upheaval (García 1978). In 1917, Puerto Ricans were granted U.S. citizenship, although they had not asked for it. With citizenship, Puerto Ricans became eligible for conscription into the United States armed forces.

The first five decades of the twentieth century saw considerable investment in sugar, tobacco, coffee, citrus fruits, needlework, military bases, subsistence farming, and a few state-owned industries. Sugar production was the predominant economic activity and occupied the best agricultural land, especially along the coasts. Intensive subsistence farming in small plots was pushed into the hills and mountains, resulting in significant soil erosion and loss of fertility. Tobacco cultivation prevailed in the eastern region and the central mountains, with environmental effects similar to those of subsistence farming (García 1978).

Aerial photographs from the 1930s to the 1950s provided by the Puerto Rico Highway Authority show a deforested landscape throughout the hills and mountains, except in the shaded coffee areas, mainly in the island's west-central region. The eastern El Yunque rainforest (Caribbean National Forest) and smaller forests under Puerto Rican government control were also exceptions to the rule. It has been estimated that only about six to seven percent of Puerto Rico was covered by forests in the 1930s (U.S. Department of Agriculture, 2000).

In 1929, at the beginning of the Great Depression, official unemployment in Puerto Rico was estimated at 30 percent (Scarano 2000). The real unemployment rate was probably much higher. In the 1930s, large numbers of Puerto Rican workers migrated to Cuba and the Dominican Republic in search of better economic opportunities. Federal economic reconstruction and relief programs began on the island in 1932 with grants from the Reconstruction Finance Administration and, in 1933, from the Puerto Rico Emergency Relief Administration (PRERA). Construction of public works with federal funds was the predominant economic activity during the Depression years.

In the late 1930s, the federal government invested in a cement plant, probably in anticipation of its participation in the war that was then brewing in Europe and Asia. In the early 1940s, military bases were built in Puerto Rico. These included Roosevelt Roads, Ramey, and Vieques Island. The Puerto Rican government later bought the cement plant and invested in the bottle, cardboard, footwear, and ceramic products industries. Funds for these projects came from excise taxes levied on Puerto Rican rum exported to the United States. In the late 1940s and early 1950s, these industries were sold to private Puerto Rican enterprises.

By 1945, Puerto Rican government representatives were already talking about the need to import private capital (technology, raw materials, administrative personnel) to make a significant dent in the very high unemployment levels. Operation Bootstrap, a development program that represented a new approach to dependent economic growth, began in 1947. A key element in the program was a ten-year exemption for corporate profits from both island and federal taxes. In the 1960s this exemption was extended up to twenty-five years for certain municipalities. Corporations paid no federal taxes until profits were repatriated to the United States. The Puerto Rican government was to provide industrial infrastructure (buildings with low rents, electricity, potable water, roads, storm drains, and sanitation systems).

The ideological assumption that informed Operation Bootstrap was that the roots of Puerto Rico's social and economic problems were overpopulation, lack of mineral and fossil fuel resources, and insufficient physical and human capital (raw materials, technology, managerial capability). Overpopulation, which was already being discussed in government circles by 1913, was treated as if it were a natural characteristic of Puerto Rico. Yet overpopulation is always related to the economic, political, technological, scientific, and cultural limits of a country. Reproduction is biological, but birth and death rates are basically sociohistorical. Overpopulation relates to the use of natural resources, but within a specific social milieu (García 1978). Eventually, foreign investment, mostly from the United States, and migration of workers to that country became the two main components of the government economic strategy.

Light industries, lured by low wages and by low-cost facilities and infrastructure, located in Puerto Rico from the late 1940s to the late 1960s. These industries were characterized by high investments in labor and low investments in machinery. The most common industrial sectors represented were apparel and textiles, metal fabrication, and electrical supplies.

To provide housing to workers in the cities and towns, a massive private and public construction effort was undertaken. Most public housing units were four stories high, but by the 1970s, structures of ten to twenty stories were being built. For the middle class, private homes and horizontal urbanization were the order of the day beginning in the 1950s, a practice that has continued up to the present. Bluntly put, this is paramount to madness, given the limited availability of land on the island, but that is what happens when the unregulated free market prevails over reason and planning.

Some of Puerto Rico's best agricultural land has been consumed by urban sprawl. Deforestation, soil erosion and fertility loss, reservoir sedimentation, disruption of potable water provision, and traffic jams have been the end results of this activity. With a series of public hearings in the Puerto Rico House of Representatives in 1966, the issues of deforestation and erosion in the mountain and coastal regions started to receive special attention, and this proved to be the beginning of concern for the environment, at least for some citizens and legislators.

Subsistence farming, along with sugar, tobacco, and coffee production, suffered significant setbacks in the 1950s and 1960s. Only meat and milk production remained relatively stable economic activities, milk more so than meat. The number of workers in agriculture decreased from approximately 229,000 in 1940 to 139,000 in 1964 and 67,000 in 1970 (Scarano 2000). Between 1940 and 1970 more jobs were lost in agriculture than were generated by industry. Commercial activities, services, and public administration recorded increases in job creation that to some extent balanced the loss of jobs in the agricultural sector (Scarano 2000).

Increasing migration began both to towns and cities in Puerto Rico and to the United States, especially to New York, and other industrial cities in the northeast. Between 1945 and 1970, 750,000 Puerto Ricans migrated to the United States; by 1970, Puerto Ricans living on the mainland numbered approximately 1.4 million. Temporary migration of agricultural workers also took place along the eastern seaboard.

Thermoelectric Plants, Oil Refineries, and Petrochemical Plants

Thermoelectric plants using residual fuel oil were built from the mid-1950s to the mid-1970s to provide electricity to industries, homes, businesses, and government entities. They became significant sources of nitrogen and sulfur oxides, and of particulate matter containing sulfates, vanadium pentoxide, nickel oxide, and polycyclic aromatic compounds. The last two of these materials have been recognized as potential carcinogens.

Acid rain, a result of nitric and sulfuric acid pollution, is another environmental consequence of these types of plants. Air pollution has been related to such conditions as asthma, chronic bronchitis, reduced respiratory capacity, prolonged and recurrent colds, and other respiratory ailments By the late 1960s, communities in Guayanilla located near oil refineries, petrochemical plants, and thermoelectric plants joined with organized workers to protest the negative health effects of these fossil-fuel-based operations. Similar protests took place in Cataño, near San Juan; these community protests continue to the present.

Four oil refineries were built between 1955 and 1971, and several petrochemical plants were built from the mid-1960s to 1971. Cheap foreign oil was imported and oil-derived raw materials were exported to the east-coast U.S. markets. In 1972, the Puerto Rican government proposed the construction of an oil "superport" to increase the importation and refining of cheap foreign oil; ultimately, though, after several years of discussion, it was not constructed. From the late 1950s to the early 1970s, the government promised to create large numbers of jobs by attracting factories that would transform oil-based raw materials into finished consumer products. In the mid-1960s, government planners promised 33,000 direct jobs and 67,000 indirect jobs. At the peak of the island's petrochemical industry in 1973, however, only 7,800 direct jobs had been generated (García 1975).

With increases in oil prices in 1973 and 1974 as a result of the Israeli-Arab war, and in 1979 and 1980 as a result of the Iran-Iraq war, Puerto Rico could no longer obtain foreign oil at a low cost. Also, the federal government's decision in 1973 to discontinue oil import quotas, which since 1965 had favored import of cheap oil to Puerto Rico, ultimately proved a fatal blow to the refining and petrochemical complexes on the island's south coast. This decision cost the refineries and petrochemical companies their competitive advantage within the U.S. market. By 1982, the oil refining-petrochemical dream had become a nightmare; most industries had closed down or were in dire straits. Direct jobs generated in these industries dwindled to around 2,000.

Yet the negative legacy of these industries remained. Oil refineries had discharged oily waste products into constructed lagoons, channels, and coastal waters, which had a devastating impact on nearby coastal fisheries. Fishermen affected by pollution spearheaded some of the earliest protests against oil refineries and petrochemical companies in Peñuelas (García 1976a, 1983a). Some of these materials, specifically solvents and gasoline components, penetrated a nearby aquifer that has yet to be cleaned up. The U.S. Environmental Protection Agency's (EPA) remediation work has been generally inefficient at this and other Puerto Rican sites affected by industrial pollution.

Nonrenewable Resources

In 1957, Kennecott Copper Corporation began modern mineral exploration in Puerto Rico. American Metal Climax followed at the beginning of the 1960s. By 1964, these firms had discovered three copper, gold, and silver deposits in the central western region (Utuado, Adjuntas, and Lares) of the main island. Additional potential deposits were discovered over approximately 37,000 acres in this mountainous, rainy, shaded coffee-growing region.

Between 1965 and the mid-1990s, opposition to open-pit mining grew, and it eventually defeated corporate and government plans. Initially, opposition to mining was based on economic and political grounds. In 1966, however, mining opponents began to develop and publicize their understanding the negative environmental consequences of mining, and they subsequently became a formidable protest movement. Citizen opposition was based primarily on the impact mining would have on soil, agriculture, water quality, human health, flora and fauna, and aesthetic considerations (García 1972).

In the late 1990s, the surface over two copper deposits was legally declared "The People's Forest," and mining was expressly prohibited there. The Puerto Rican government bought the land and established a joint management arrangement between the Commonwealth and municipal governments. Additional efforts have been made more recently to add significant tracts of land to the original forest, to create a continuous system of protected areas in the Cordillera Central.

During the late 1950s, mining companies also undertook preliminary exploration of nickel, cobalt, and iron deposits in western Puerto Rico. Expansion of the exploration and possible exploitation of these mineral resources was discussed in the 1970s. Opposition to extensive surface mining arose, based on economic, political, and environmental grounds, and this mining proposal was shelved in the early 1980s (García 1984a).

A third attempt to exploit nonrenewable resources began in 1975. Information surfaced in the media that Mobil and Exxon were interested in exploring significant seabed and land areas in northern Puerto Rico for the possible presence of natural gas and oil. This interest had been piqued by site investigations undertaken in conjunction with plans for constructing nuclear power plants on the island. Geophysical studies had revealed the presence of potential hydrocarbon deposits. Puerto Rican public opinion crystallized around the demand that the island government, not the corporations, finish the exploration and then negotiate the exploitation of hydrocarbon deposits, if any expansion of the dry well resulted. Lingering suspicions remain as to the extent of Puerto Rico's natural gas and oil deposits.

Pharmaceutical Corporations

In the 1960s and the 1970s the Puerto Rican government made a significant effort to attract pharmaceutical companies to the island. During the 1960s, this investment took place mostly in end production of medicines, specifically preparation of tablets, emulsions, and solutions, and in packaging. By the end of the 1960s, Puerto Rican plants had begun to synthesize medicines through chemical and biochemical processes. This sector of the drug industry expanded dramatically for almost three decades, and Puerto Rico became one of the most important centers of pharmaceutical production in the world.

Section 936 of the Internal Revenue Code favored expansion of the pharmaceutical sector. This regulation allowed parent companies to repatriate virtually tax-exempt profits from subsidiaries in Puerto Rico to the United States. Their only requirement was to pay a small "tollgate tax" on profits to the Puerto Rican government. The transfer of profits from parent companies to Puerto Rican subsidiaries through raw materials purchases, product prices, patent controls, and other creative accounting schemes became widespread, until the U.S. Congress decided to eliminate Section 936 over a ten-year period, starting in 1996. The pharmaceutical and chemical companies were notorious for inadequate disposal of hazardous waste during the late 1960s and mid-1970s. Disposal took place in inappropriately constructed lagoons, in sinkholes, in channels and rivers, and in domestic waste landfills with no adequate safeguards.

Electric machinery parts companies were also a source of soil and water pollution during this period. During the 1970s and early 1980s, solvents and heavy metals were found in landfills, wells, and surface water in northern, southern, and southeastern Puerto Rico. The sources of these pollutants were traced to the above-mentioned industries, as well as to thermometer producers, oil refineries, and petrochemical plants (Torres-González and Wolansky 1984; Guzmán-Ríos and Quiñones-Márquez 1985).[1]

Vocal opposition from community groups arose at that time to the location of these industries and to the permitting process that allowed them to discharge waste into rivers, landfills, and public treatment plants. Demands to clean up wells, landfills, and soil and to protect the health of those affected were common. Protests around the ineffective operation of domestic waste landfills were also the order of the day. At least ten sites where wastes were improperly disposed of were eventually declared Superfund sites and placed under EPA jurisdiction.

Puerto Rico's growing electricity needs have been an area of great contention. Between 1968 and 1976, there were numerous demonstrations in Puerto Rico protesting the proposed construction of nuclear-powered electric generating plants. In 1977, nuclear power was eliminated as an option for electricity generation. Between 1979 and 1997, opposition arose to proposed construction of three coal-fired electric generating plants. Only one of these projects (AES) reached completion—a plant built in southern Puerto Rico. After three and a half decades of searching for strategies and solutions to meet energy needs, the government has supported construction of natural gas and distilled fuel oil plants. This has meant a significant reduction in consumption of power generated by thermoelectric plants that used residual fuel oil with high sulfur and asphaltene content. This victory has come after years of protest, spearheaded by Cataño residents, against the Puerto Rico Electric Power Authority (PREPA) and EPA. Ultimately energy conservation improvements in the existing electricity generating system, along with use of renewable energy sources, must be part of the solution for Puerto Rico's long-term electricity needs (García 1993).

Military Activities

For more than sixty years, a significant portion of Puerto Rico was used by the U.S. armed forces for military base locations and operations. Poverty, prostitution, and pollution were the main products of this military occupation. In the 1950s human life was made intolerable by drunken sailors' abusive behavior toward women and old men. Nowhere were military activities more destructive than on Culebra and Vieques islands, to the east of the main island. More than a thousand acres in eastern Vieques were used to practice with napalm, uranium-tipped bullets, and bombs of up to 2,000 pounds. Navy maneuvers destroyed coral reefs, lagoons, flora and fauna; the roar of airplanes and bomb explosions made life miserable for residents. When the Navy left the bombing range in Culebra in 1975, it simply increased bombing practice in Vieques. Cancer rates and fatalities skyrocketed among Viequenses, beginning in the 1980s.

After decades of struggle against Navy control of more than two thirds of Vieques, a stray bomb from a combat airplane killed a civilian guard working for a private contractor at a military post. This incident led to four years of renewed struggle. Eventually the Viequenses succeeded in ousting the Navy from their land, but most of the land was transferred to the U.S. Department of Interior.

Now, Viequenses and other Puerto Ricans are continuing the fight for cleanup, control of their land, and true economic and social development (McCaffrey and Baver, in this volume).

The 1980s to the Present

Economic Changes

Since the 1980s, there has been a marked shift in the Puerto Rican economy to banking, finance, insurance, large commercial centers, communication, education, tourism, and physical infrastructure construction (Gutiérrez 1996; Asociación de Industriales de Puerto Rico 2004). Horizontal urbanization has continued for both residential and commercial activities. Due to the scarcity and high cost of urban land, construction of four-story, walk-up apartments has increased in the San Juan metropolitan region. The pharmaceutical, electronics, and medical products industries expanded until the early 1990s, and they improved their waste disposal practices. Yet, some industrial waste is still sent to regional water treatment plants, creating problems for their operation. Domestic solid waste disposal has improved somewhat, but there is still a long way to go. Food canning, although with increasing import of raw materials, remains an important economic activity.

During the same period, apparel, metal fabricating, and textile and shoe manufacturing declined and most firms in these sectors left Puerto Rico for more profitable locations. Their economic viability became even more questionable with the signing in 1994 of the North American Free Trade Agreement. NAFTA made Mexico a more profitable location for these firms because of lower salaries, lower transportation and electricity costs, and weaker environmental controls.

Under the Caribbean Basin Initiative, first approved by the U.S. Congress in 1983 but later reauthorized, countries like the Dominican Republic were favored as locations for light industry because of their lowered trade barriers with the United States. Cheap labor in Asian countries has also played a role in displacing this type of industry from Puerto Rico. Although the figure for industrial jobs in Puerto Rico hovered around 160,000 for many years, the last decade has seen a marked reduction in industrial jobs as a result of the CBI, the North American Free Trade Agreement (NAFTA), and the economic calculations of transnational corporations. Economic, technological, educational, and other social transformations in Puerto Rico have also contributed to changes in the job categories for Puerto Rican workers.

Socioeconomic Conditions

Census 2000 data indicate that average official unemployment in Puerto Rico hovers around 19 percent (Departamento de Salud 2000). While unemployment levels around of around 14 percent are typical for the San Juan metropolitan area, in certain municipalities they range between 24 and 29 percent. 46.2

percent of the city's 3.9 million inhabitants lived below the federal poverty line (U.S. Census Bureau, *Population Census 2000*). Households headed by women constitute 27 percent of families in Puerto Rico. 71 percent of these families live below the poverty level, with an average income of $6,888 (Millán-Pabón 2003).

A study done by the Commonwealth Office of Special Communities revealed that in 686 communities, 28.9 percent of inhabitants neither work nor are looking for a job. More than half a million Puerto Ricans live in these 686 "Special Communities." 46 percent of residents in the Special Communities aged twenty-five years or older had not finished high school, and a majority are more dependent on government health care than on government food checks (*El Expreso* 2004). Specifically, 64.2 percent of households have the governmental insurance health plan, while only 32.2 percent are included in the Nutritional Aid Program (Programa de Asistencia Nutricional). Since almost one fifth of residents (19 percent) are sixty years or older, they receive federal Medicare or Medicaid benefits.

Among prevailing general social conditions in Puerto Rico, high levels of child poverty and school dropout rates are notable (Kearney, 2003). A study done under the auspices of the National Council of La Raza found in 1999, 58 percent of children lived in families whose income placed them below the federal poverty limits (Figueroa 2003). Puerto Rican children are three and a half times more likely to be poor than children in the fifty states of the United States. Rural municipalities in Puerto Rico with less than 10,000 inhabitants, such as Vieques, Maricao, and Las Marías, have the highest levels of child poverty, with rates of 81 percent, 77 percent, and 76 percent, respectively (Figueroa 2003).

One in seven adolescents between the ages of sixteen and nineteen do not finish high school. This equals a 14-percent youth dropout rate compared to 10 percent in the United States. In the following municipalities, at least one fifth of adolescents are school dropouts: Adjuntas and Aguadilla (22 percent), Luquillo (21 percent), Ciales, Guánica, and Vieques (20 percent). Very high levels of unemployment prevail in this young population, and the sale and use of illegal drugs is rampant.

Public Services

There are around 2.5 million registered cars in Puerto Rico for close to 4 million inhabitants. Traffic jams are the order of the day, even if there are more miles of road per square mile than in most countries. Many workers lose between two and four hours every day driving their children to and from school and going to and from work. Enormous amounts of gasoline and human energy are wasted in this absurd transportation madness. Particulate matter containing aromatic polycyclic hydrocarbons, carbon monoxide, nitrogen oxides, and unburned and partially burned hydrocarbons are spewed into the air. Along with emissions from thermoelectric plants, pharmaceutical and chemical companies, cement plants, and construction activities, auto emissions are a significant factor in the continuing and accelerating decline of human health and the general environment.

The transit nightmare is related to massive construction for both residential and commercial purposes. In the year 2000, the United States had eighty persons per square mile; Australia had eight and Puerto Rico had 1,112. Less than one third of Puerto Rico is flat land, yet a very high percentage of flat land has been consumed by horizontal urbanization, which has caused increasing deforestation, soil erosion, reservoir sedimentation, flooding, and reduction in agricultural and forest cover over the past four decades (UMET 2001). At present, there is more forest coverage than in the 1930s, but urban sprawl continues to threaten this gain (U.S. Department of Agriculture 2000).

Potable water service is very poor or almost nonexistent in many low-income communities located in hilly or mountainous terrain. Water scarcity is even worse during dry spells, like the one in 1994–1995 that affected the San Juan metropolitan region. Approximately $500,000,000 has been invested in the construction of a filtration plant and a forty-mile aqueduct to provide additional potable water to this region. However, poor communities in other regions still suffer from an almost perennial scarcity of clean drinking water.

Approximately 45 percent of Puerto Rican houses lack access to sanitary discharge systems and wastewater treatment plants. Most of these are located in poor communities outside the city limits. Discharge of wastewater into creeks and rivers affects surface water quality. When this water is used in filtration plants that operate improperly or above capacity, the result for humans can be gastroenteritis and infectious diseases.

Puerto Ricans generate 8,000 tons of domestic solid waste, or around four pounds per capita, per day (Autoridad de Desperdicios Sólidos 2003). This figure does not include construction debris and hazardous industrial wastes. 90 percent of such waste is disposed of in thirty-one landfills. Of these, twenty-seven do not comply with the EPA and Puerto Rican solid waste regulations in force since April 1994. Less than 2 percent of potentially recyclable solid residues are incorporated into new production processes in Puerto Rico (García 1998, 2002).

Poorly operated landfills are the source of pollutants that contaminate surface and underground water. The presence of organic and inorganic pollutants in surface water leads to profoundly negative effects on aquatic life. Additionally, polluted underground water cannot be used as drinking water for humans and other animals.

In sum, many of Puerto Rico's infrastructure challenges and some of its pollution problems stem from its choice to urbanize horizontally. To date, no attempt to modify the island's tax code to favor vertical construction and living has been successful. Nor have zoning regulations been used successfully to stem the tide of urban sprawl. Puerto Rico's construction industry, which is promoting sprawl, is enormously powerful politically. The temporary jobs the industry provides can be important factors in determining who wins many island elections. It is therefore very difficult to protect the environment and to control urban sprawl in the context of the high levels of unemployment that persist in Puerto Rico.

Conclusion

Countries that import capital also export economic surplus in the form of profits controlled by industrial, commercial, and banking enterprises. Thus, these dependent countries lack the capital they need in order to meet their educational, health care, recreational, solid waste disposal, wastewater treatment, potable water, public transportation, cultural, research, and technology requirements.

Economic dependency gives birth to relative overpopulation, which is at the heart of many of the social ills so prevalent in Puerto Rico and, to some extent, among Puerto Ricans in the United States. More than 40 percent of Puerto Ricans live in the United States; many of these share economic, educational, health, and cultural problems similar to those in Puerto Rico. In addition, many still suffer from language barriers and racism.

Migration patterns to the United States, however, have become more complex. During the past three decades, a high percentage of Puerto Rican migrants have been economically active or retired professionals; this translates into a significant loss of public investment in education on the island. The brain drain in a country struggling to solve pressing economic and social problems presents a painful contradiction.

Many Puerto Ricans have excelled in sports, arts, music, acting, engineering, scientific research, and other avenues of educational achievement. Yet others have done less well because economic, educational, and cultural opportunities are still highly unequal. Puerto Ricans will have to make difficult decisions in the years to come. We hope these decisions are made wisely and collectively, to accomplish justice with freedom for all. Both the social and the natural environments are at stake.

Note

1. Carlos E. O'Neill, of the EPA, said in a 1983 personal communication that the U.S. Geological Survey tested the wells in 1983 and found volatile organic compounds in the Ponderosa well.

CHAPTER 7

Seeking Agricultural Sustainability

CUBAN AND DOMINICAN STRATEGIES

BARBARA DEUTSCH LYNCH

Introduction

Cubans and Dominicans have suffered from a perverse pattern of environmentally destructive export agriculture. This pattern, established during the colonial period, has been unstable because the land and water management practices associated with export production have exacerbated soil erosion, salinization, and pest problems, and because dedication of prime lands to export agriculture has driven domestic food production to forested, steeply–sloped, and fragile lands. In 1990 the prospects for an environmentally sensitive agriculture appeared bleak despite the proliferation of small sustainable-agriculture projects, but by 2003 a transition toward environmentally friendly food production appeared to be under way.

In the past two decades, myriad development programs—large and small, international and national—have been launched with the goal of making agriculture more sustainable. Sustainability is a utopian and inherently fuzzy concept. In its broadest form, this concept suggests the development and diffusion of ecologically sound agricultural technologies and the creation of a social, economic, and political environment supportive of food production. Narrowly conceived, it refers to a set of land management practices including erosion control, careful water application, and integrated pest management to reduce agrochemical use.

In this chapter, I argue for a definition of sustainability that emphasizes food production within the agricultural sector, preservation of agricultural landscapes, and cultural validation of food production and producers. This definition suggests the need to look beyond particular farming practices to the economic, social, and political factors that influence the generation and legitimation of agricultural knowledge and the allocation of land and other resources among the competing uses and within agriculture. Specifically, we need to examine the relationship between smallholder food production and "modern" production on prime lands, the relationship between agricultural production and national responses to global economic pressures, and land use competition. We also need to understand how national cultures support or denigrate food producers and environmentally friendly forms of agriculture.

Sustainable agriculture programs in keeping with this broader vision would include design and management of agro-ecosystems so that they remain productive over time,[1] prioritization of food production over competing uses of agricultural land, preservation of the existing stock of arable land from conversion to industrial and residential use, and cultural validation of food producers and agro-ecosystems geared to subsistence and local market production. Without a crystal ball, we cannot know whether this kind of programmatic emphasis would allow populations to meet their food needs over time. For this reason, I will refer to agricultural policies and practice that fall under this umbrella as "alternative" agriculture.

Some nations are not overly concerned with the preservation of land for domestic food production. Rather, they rely on imports for food security. But people need to eat, and if imported food is unaffordable they will do whatever they need to do in order to feed their families. If the land base for food production is inadequate, it will be overexploited, leading to the erosion and deforestation that have aroused international concern. Conversely, in low- and middle-income countries, preservation of the agricultural land base is closely tied to food security, environmental security, and national viability.

Increasing food production on a constant or shrinking land base requires intensification either in the form of increased labor and management or increasing use of agrochemicals and water. The former strategy requires recognition of the positive contributions of locally adapted agricultural forms; the latter requires a "level playing field"—optimal conditions for the growth of high-yield varieties (Ploeg 1990). The latter model, called "conventional" or "modern" agriculture, has proven unsustainable, even over relatively short time horizons,[2] but it carries weight in Caribbean agricultural research and development circles. The economic limits of conventional export agriculture became apparent with declining commodity prices, and with the rising cost of imported seed, pesticides, and fertilizers on the one hand and increased demand for organic coffee, cacao, fruits, and vegetables on the other.

To what extent has alternative agriculture gained respectability among agricultural policy makers in Cuba and the Dominican Republic? Does respectability imply a transition to an environmentally and culturally sound agriculture? The two countries have had similar histories up to a point, but different approaches to environmental protection, food production, land use, and land tenure, and very different incentives to adopt alternative strategies. In this chapter, I ask whether alternative agriculture experiments represent a reversal of past patterns and the beginning of a transition toward sustainability. I begin with a review of the obstacles to agricultural sustainability in both countries. I then outline steps that each country has taken to achieve the elusive goal of sustainability, compare the two strategies, and assess the potential of each for national environmental security.

Obstacles and Alternative Agricultures

Caribbean agrarian societies were shaped by highly extractive, export-oriented plantation economies and by the semi-clandestine agriculture that evolved on its edges.[3] This agriculture, associated with Tainos, maroons, and poor peasants, accounted for much of Caribbean food production. It was characterized by polyculture with an emphasis on root and tree crops and often by shifting cultivation. The term *conuco* is often used to refer to these food production systems as well as to the plots devoted to food production. Spanish conquest brought with it the introduction of Old World crops and livestock, the institution of slavery, and the decimation of local Taino populations, which occurred early and had immediate impacts on local food systems. Although plantation agriculture established a foothold on both islands in the sixteenth century, before the nineteenth century it constituted a significant agricultural sector only in the French colony of Ste. Domingue (later Haiti), the most important sugar producer in the Caribbean. The eastern two-thirds of Hispaniola (what would become the Dominican Republic) was devoted to cattle and food production.

The Cuban sugar economy grew in importance following the Haitian Revolution, but in the early nineteenth century the plantation sector in Cuba, as in the Dominican Republic, left a land base more than adequate for food production. By the late nineteenth century, expansion of sugar production onto flat, fertile lands marginalized food production and food producers (Barnet 1994; González 1993; Franks 1997; Baud 1987). This pattern continued into the late twentieth century. In the 1960s and 1970s, mechanized rice production was introduced on large state farms and agrarian reform enterprises in both countries, while production of Caribbean dietary staples—yuca, yautia, ñame, plantains, pigeon peas, beans—was relegated to small farms, often on infertile or steeply sloped lands.

After 1992, the Cuban state selectively relinquished control over major productive sectors to stimulate satisfaction of domestic demand (Deere 1993, 1995; Torres Vila and Pérez Rojas 1995; Enriquez 2003). The 1992 Cuban agrarian reform included the transfer of state-farm sugar lands to newly created Basic Units for Peasant Production (Unidades Basicas de Producción Campesina [UBPCs]), but before 2000, both the Cuban and the Dominican governments exercised control over sugar and rice production, and in both countries the balance between conventional agriculture and local food production remains problematic. Plantation agriculture would become less rational with the decline in prices for tropical commodities, and policies based on food and feed imports were unlikely to be viable over the long run for countries lacking foreign exchange. Moreover, cultural denegration of *conuco* production as backward, rustic, and African contributed to a systematic undervaluation of its contribution to Caribbean economies.

Inattention to food production in the Dominican Republic and in Cuba up until the 1990s had the same impact—increasing the total amount of cultivated land and pushing food production to fragile lands. The two countries' paths diverged in the 1990s as Cuba recognized the need for local food self-sufficiency and took measures to stimulate food production. The result was a "repeasantization" of Cuban agriculture—the reemergence of labor-intensive food production on small plots (Enriquez 2003). Ability to sell a portion of their production at high market prices meant that small Cuban producers enjoyed a greater degree of economic security than their Dominican counterparts.

Dominance of the Green Revolution Paradigm

The potential of alternative agriculture for meeting world food needs is still a subject of lively debate. However, it is widely acknowledged that agricultural modernization associated with the Green Revolution of the 1960s has contributed significantly to the instability of agriculture in the tropics. Agricultural modernization is best described as the externalization of different factors of production—the application of outside knowledge, fertilizers, seed, water, and pesticides—to achieve uniform and predictable results (Ploeg 1990). In practical terms this has meant producing uniform seeds—whether through conventional breeding practices or through genetic modification—that outperform local varieties under optimal conditions of soil fertility, pest control, and moisture. The Green Revolution also entailed creation of a social apparatus for knowledge generation. In principle, knowledge would be acquired through carefully controlled experiments conducted at research centers, universities, and experiment stations. Information about optimal soil conditions and pest control strategies for particular crops would be disseminated through national extension networks, often funded by international donor agencies (Levins and Lewontin 1985). Adoption of recommended packages and cultivation plans became a prerequisite for obtaining credit.

Until 1986, this conventional agricultural strategy predominated in Cuban institutions and in the international assistance projects undertaken to modernize Dominican production. The technological package of improved seed, fertilizer, and pesticides became the basis for lending to small farmers on agrarian reform settlements in the Dominican Republic. Emphasis on high-yielding, high-input, experiment-station-designed cropping systems had several negative impacts. First, local knowledge about climate, soil structure, and pest conditions in the design of production strategies was underutilized. Second, small farmers on reform settlements enjoyed little freedom to experiment with alternative methods. Third, locally based knowledge and low-input practices were deemed "unscientific." This made it more difficult for agricultural researchers to evaluate, let alone recommend ecologically sound agricultural practices that had been developed *in situ*. It also bolstered the position of the petrochemical industry.[4]

Rural Land Tenure

Land concentration, lack of access, and tenure insecurity can be major obstacles to agricultural sustainability. Land tenure in and of itself would not be expected to have significant environmental implications, but where land ownership is highly concentrated, incentives for monoculture are strong, and, historically, crops on large holdings—with the notable exception of rice—were destined for export rather than for local consumption. Second, both traditional plantation and modern export agribusiness economies reduced access to land for polycultural root-crop farming (*conucos*). Equally important, preservation of these systems was a very low priority in national agricultural strategies. Third, because managers of large holdings, whether Dominican growers or Cuban state employees, tended to define efficiency in terms of yield per unit of labor, they used machinery, petrochemical inputs, and water in ways that were neither economically nor ecologically sound. Fourth, with enclosure of prime lands, food producers have been forced to cultivate on infertile or steeply sloped land or to shorten fallow cycles so that land cannot recuperate. Where tenure is insecure, farmers will be reluctant to adopt land conservation practices that are costly or labor-intensive.

Land concentration was extreme in Cuba and the Dominican Republic in the 1950s. Both countries carried out land reforms beginning in the early 1960s, but in neither case did these reforms produce a nation of small, independent food producers. On the eve of the Revolution the 8.5 percent of Cuban farms over 403 hectares in size accounted for 73.3 percent of agricultural land (Zimbalist and Brundenius 1989). Holdings of less than 67 hectares (68.3 percent of farms) accounted for only 7.4 percent of agricultural land. After the 1959 revolution, the Cuban government expropriated holdings larger than 1,000 acres, leaving about 25 percent of agricultural lands in the hands of large producers, many of whom were hostile to the revolution (Evanson 1994). The 1963 Reform Act eliminated the large-scale private sector, leaving 60 percent of agricultural land in state hands and 30 percent in the hands of small farmers. By the 1980s the small producer sector (individuals and cooperatives) accounted for only about 18 percent of agricultural land (Diaz and Muñoz 1994). State-run enterprises continued to occupy much of the nation's prime land well into the 1990s, but a significant shift toward farmer ownership and management occurred after 1989, as the Cuban government sought to achieve food security in an increasingly hostile economic environment. A land reform enacted in 1993 reduced the state's holdings to just 27 percent of Cuba's cultivable land in 1995 (Economist 1996). Small individual holdings, agricultural cooperatives, and the new UBPCs accounted for the remainder. Also, during the 1990s, many Cubans became farmers for the first time. While the rise of the small farm sector has increased pressure on hill lands, particularly in the Oriente, this may be out-weighed by the closer ties Cuban cultivators now have to the lands they till. In sum, revolutionary legislation guaranteed all but the largest Cuban food pro-ducers access to land and tenure security.

In the Dominican Republic, land concentration forced expansion and later intensification of food production on marginal lands to meet local food needs. The land colonization schemes of the Trujillo era provided some campesinos with holdings sufficient to supplement low-wage and part-time labor on planta-tions and in industry (Turits 2003). Ten years after Trujillo's death in 1961, 1.01 percent of farms occupied 47 percent of the nation's agricultural land (Dore 1982). In 1960, of the nation's estimated 5,750,560 farmland acres, 452,519 were producing sugar for industrial sugar mills. The Dominican agrarian reform cre-ated small holdings on irrigated settlements or *asentamientos*, but did not sig-nificantly shift the balance of ownership in favor of small farmers. Census data from 1981 show that 1.83 percent of landowners owned 55.2 percent of culti-vated lands, while 81.7 percent of landowners accounted for 12.1 percent. In addition, 409,959 peasants were landless (del Rosario et al. 1996, 7). Many others lacked secure title to the fields they cultivated. The effects of landlessness and near landlessness were migration to the nation's cities, increasing reliance on off-farm income, and a shift away from labor-intensive production.

In sum, land concentration remained a problem in both countries until the 1990s, although access to land and tenure security were serious concerns only in the Dominican Republic. Until quite recently, both countries featured large state sectors dominated by sugar production. In Cuba, increasing empha-sis on food security, the growing profitability of food production with the re-opening of farmers' markets, and the introduction of the UBPCs in the Cuban state farm sector in the 1990s led to an increase in the number of small owner-operated units. In contrast, in the Dominican Republic the declining contribu-tion of export agriculture in a neo-liberal context provoked a general exodus from agriculture.

Withdrawal of Land from Agriculture

Perhaps the major continuing threat to agricultural sustainability is with-drawal of land from agricultural use. In rural and peri-urban areas, privatization and land speculation continue to erode the land base for crop production. The agricultural landscape itself is urbanizing as farms give way to outdoor facto-ries for water and agrochemical-intensive pineapple, citrus, and pork produc-tion. Farm workers are housed in barracks or in urban settlements without amenities or basic services. The Dominican Cibao, a rich agricultural region stretching from La Vega to Santiago, has become a metropolitan area. The state and private landowners have converted agricultural lands into industrial parks, and new neighborhoods for workers have arisen around them. Santiago's newly constructed airport occupies highly productive lands near Moca. Additional with-drawals of cultivable lands have taken place with construction of suburbs and highways. Even *conucos* on the slopes of the Cordillera Septentrional are being converted into vacation home sites. As the Cibao has urbanized, we have seen on the one hand a shift from mixed-crop production to low-labor monocultures like plantains and yuca, and on the other hand very intensive production of high value crops such as tobacco and landscaping plants. Since 2000, even the Cibao's prime tobacco lands are being turned into urban settlements. Peri-urban development has also damaged the Nigua watershed, where gravel mining for urban construction competes with small-scale agriculture and grazing. In the Cibao and Yuna valleys, urbanization and industrialization have reduced and degraded the water supply for irrigated agriculture, resulting in further withdrawals of land from productive use.

International migration has also accelerated withdrawal of agricultural land from productive use in the Dominican Republic. Lacking access to their children's labor, aging farm families often adopt cropping strategies that minimize labor inputs. For example, in the mid-1990s, binational residency and price volatility in global markets encouraged a shift from coffee to pasture production in the Cibao.

Ironically, protected area designation has had a perverse effect on Dominican agriculture. Traditionally, Dominican food producers practiced shifting cultivation as a way to allow soils to recuperate. In the 1990s, the Dominican government extended protected area status to remote areas, closing them to cultivation and thereby increasing pressure on lands outside the reserves. The area from which cultivators are excluded has grown rapidly. In 1988, national parks and protected areas accounted for 11 percent of the Dominican land mass. By 1994, that figure had increased to 21.5 percent (World Resources Institute 1996, 241).[5] A July 1996 presidential decree established six new parks, extended the boundaries of another, and created six scientific and biological reserves and four natural monuments, bringing the total protected area to more than 30 percent of the national territory. Large-scale tourist resorts have occupied public lands, withdrawing them from agricultural use. By reducing the land base for shifting cultivation, these enclosures have contributed to the shortening of fallow cycles and the overuse of fragile lands, and capricious eviction policies associated with protected area management have discouraged small producers from investing in tree crops or long-term conservation practices on lands to which they do not hold secure title.

By the 1990s, withdrawal of land from food production contributed to a decline in Dominican food self-sufficiency In 1976–77, food imports accounted for 17 percent of consumption. This figure climbed to more than 50 percent in the period 1982–86 (CEUR 1993). The proportion of uncultivated "cultivable" lands declined from 13 percent in 1977 to 6 percent in the early 1990s, leading to an environmental squeeze on small producers who had to invest more labor for every unit of output in a situation where labor was increasingly scarce (CEUR 1993). And the number of hectares per capita has fallen from 0.23 in 1983 to 0.19 in 1993 (World Resources Institute 1996, 241)

The loss of agricultural land has been less significant in Cuba. Protected areas account for less of the national territory,[6] and their management includes attention to the needs of resident cultivators. The amount of land cultivated in mixed root crops increased by 60 percent from 1958 to 1989. For the same period, rice lands grew by 52 percent, lands planted in beans by 35 percent, and the amount devoted to other vegetables increased by more than a factor of 20. Much of this increase may have come at the expense of corn production, which decreased by 47 percent. This shift in land to food crops was probably facilitated by feed imports in the period before 1989 (Deere 1993). Cuban emphasis on extending urban services to small towns and rural areas may also have contributed to the preservation of the agricultural land base. The main threats to preservation of cultivable lands are tourism and peri-urban growth around Havana.

Caribbean Agriculture in the Global Economy

Until very recently, the terms of Cuban and Dominican insertion into the global economy have limited the potential for sustainable agriculture in both nations. Structural adjustment and trade agreements based on comparative advantage have deepened Caribbean dependence on agricultural exports and reduced the resources available to national governments for research and extension in support of local food production.[7]

Export-driven production strategies are becoming more sustainable, but throughout the twentieth century, they relied heavily on conventional technologies, leading to salinization, erosion, and the loss of crop-plant and livestock biodiversity. In the 1990s, prices for leading tropical commodities, except for bananas, fell steadily. Sugar took the sharpest dive, followed by cacao, coffee, and rice (World Bank 1993; also see Raynolds 1994). Despite this trend, emphasis on sugar as a source of foreign exchange continued in both countries throughout the 1960s and 1970s (Enriquez 1994; Ghai, Kay, and Peek 1988; Zimbalist and Brundenius 1989) and persisted in Cuba until 2002 (González 2002; Sinclair and Thompson 2001).[8] Conversion of land to sugar production was closely associated with deforestation. Industrial sugar mills (*ingenio* or *central*) required fuel and wood for the rail lines that linked cane fields to the mill. In the Dominican southwest and Los Haitises regions, timber extraction for cross ties was a major factor in deforestation. Cuban sources note that due to expansion of the U.S.-controlled sugar economy from 1900 to 1959, 75 percent of Cuba's forest cover was lost (Ministry of Science, Technology and Environment 1995, 48). However, despite a 71 percent expansion of the acreage devoted to sugar from 1958 to 1998 (Enriquez 1994), a net increase in forest cover is reported (Ministry of Science, Technology and Environment 1995, 48). If accurate, these data would indicate that while expansion of sugar production historically entailed deforestation, this association may not hold as the sector matures.

Whatever its effects on forest cover, it is difficult to reconcile sugar production with food security. In his 1952 "History Will Absolve Me" speech, future president Fidel Castro saw dependence on sugar cultivation as a barrier to full employment and food security, and immediately after the revolution attempts were made to diversify production. However, by 1963, sugar again became Cuba's primary source of hard currency. While food production increased during the 1970s, until 1990, more than half of Cuban agricultural land was dedicated to export production and more than 50 percent of Cuban caloric intake came from imported sources (Trueba González 1995). The large state farm sector (63 percent of agricultural land) featured highly mechanized production, whether of sugar, rice, or potatoes. By the late 1980s, Cuban sugar production was based on new high-yield stock; land preparation and most harvesting was mechanical, 75 percent of sugar plantations used herbicides, and 30 percent of sugar lands were irrigated (Zimbalist and Brundenuis 1989; Figueras 1992).

The fall of the Soviet Union and dissolution of COMECON (Council for Mutual Economic Assistance) caused a profound economic crisis in Cuba, referred to as "The Special Period in Time of Peace." The nutritional impacts of the crisis were severe, and they caused a shift in agricultural sector priorities in favor of domestic food security. As Enriquez (1994) notes, because root crop and bean cultivation was resistant to mechanization, its production was concentrated on small cooperative and individual holdings, so Cuba's new Food Program focused on small producers and cooperatives.[9]

Despite its commitment to food self-sufficiency, Cuban emphasis on sugar production continued during the Special Period. Economists and government officials insisted that however important the food program, it must be achieved in a way that "assure[s] the continued production of our principal source of export [earnings]: the sugar agroindustry" (Ministry of Science, Technology and Environment 1995, 2). Simultaneous pursuit of the conflicting objectives of increasing exports and domestic food production required that Cuba dedicate more land to agriculture and make production far more efficient in both sectors. The sugar sector employed some 400,000 workers (Sinclair and Thompson 2001). In 1990, 1.8 million hectares were devoted to cane; this figure dropped to 1.4 million hectares in 2000 (Sinclair and Thompson 2001). In 1995–1996, the Cuban government borrowed heavily to purchase the petrochemicals and spare parts for farm machinery on which industrial sugar had come to depend (*Economist*, April 6, 1996). In the short run, the gamble appears to have paid off by increasing yields, but the terms of trade continued to deteriorate.

Cuba's continuing reliance on sugar in the 1990s despite declining prices on the world market illustrates the importance of agriculture for foreign exchange generation. Only in late 2002 did the government risk major cutbacks in sugar production; it announced that "60 percent of existing sugar fields would be given over to other agricultural production and that former mills would be converted to food processing plants" (González 2002). As sugar's profitability waned, tobacco and citrus exports assumed greater importance. Tobacco, a short-season crop easily alternated with food crops, has become very pesticide-intensive, pos-ing severe health risks for agricultural workers. Citrus cultivation too has typically been pesticide dependent in order to meet the aesthetic requirements of importers, and its input-intensive cultivation is centered on the Isle of Youth.

In the Dominican Republic, the amount of cultivated land dedicated to export production tripled from 1950 to 1995 and grew by nearly a third from the mid-1980s to the mid-1990s, hampering the nation's ability to meet domestic demand for crops like beans, bananas, and root crops (del Rosario et al. 1996, 77–78). With the collapse of the sugar boom and the Caribbean Basin Initiative, emphasis shifted to winter fruit and vegetable exports, tomatoes, and oil palm. U.S. and Israeli technical assistance teams and private enterprises promoted tomatoes, melon, oriental vegetables, and pineapple on irrigated settlements. At the same time, with minimal support from the Secretariat of State for Agriculture (SEA), production of local root crops for international markets took off (Lynch 1992; Geilfus 1985). But on the whole, SEA, the Consejo Estatal de Azucar (CEA), and the Institute for Agrarian Development (IAD) demonstrated little interest in domestic food crops other than rice, and to some extent beans.

By the late 1980s, export production began to replace domestic food production even within the *conucos*. In Zambrana-Chacuey, small holders grew pineapples for export; and in the late 1980s and early 1990s, shifting cultivators in Los Haitises grew the root crop yautía for U.S. and Puerto Rican markets. Cultivators on agrarian reform settlements raised export crops under contract: agribusiness enterprises typically advanced credit on the condition that farmers use it to invest in a technological package of improved seed, fertilizer, and pesticides and that they sell their crop to the enterprise. Contract production removes decision making from farm field to the agribusiness firm and increases dependence on imported inputs, technologies, and seed stock. In the late 1980s, the contract production strategy backfired with a vengeance when multiple cropping and heavy water applications provided an atmosphere favorable to the proliferation of insect pests (Murray 1994). Widespread application of broadspectrum pesticides in nontraditional vegetable production methods had led to infestations of *Thrips palmi* in La Vega's oriental vegetable areas and white fly in the Azua tomato fields. The response to these epidemics was increased application of pesticides, to the point where the U.S. Department of Agriculture halted imports of melons and oriental vegetables (Raynolds 1994). As prices and production levels fell, farmers were left with loans they could not repay.

With declining sugar prices and the white fly crisis, the contribution of export agriculture to GDP declined. Also in the 1990s, Dominican coffee growers faced increasing competition from Vietnam, and, with elimination of trade barriers, high-quality Caribbean bananas had to compete with cheaply produced bananas from Central America and Ecuador (Fireside 2002; Raynolds 2000). Coffee and banana producers found that they could survive more easily in the global economy by producing for organic and fair trade markets, and by 2003 the Dominican Republic had become the region's foremost exporter of organic bananas.

Cultural Obstacles to Agricultural Alternatives

In the Dominican Republic, cultural obstacles to the shift to alternative agriculture were substantial. Even more severe were the barriers to polyculture and shifting cultivation. Official statements on appropriate uses of the public domain, shaped by the culture of export agriculture and the rhetoric of Dominicanization, generally denigrated small cultivators. Their contribution to GDP was systematically undervalued, and they were routinely blamed for deforestation, dam siltation, and other environmental problems. Anti-*conuco* rhetoric—which appears as late as 1997 in the environmental literature—had two components. The first was the anti-Haitianism used since the Trujillo era to create a fixed but highly permeable border and to control the movements of Haitians within the Dominican Republic. *Conuco* cultivation was linked to "African (Haitian) backwardness," maroon subversion, and in the 1980s with environmental destruction. A second strand of anti-*conuco* discourse was environmental, and was associated with watershed protection and dam building. Blame for deforestation, erosion, and high dam siltation rates was assigned to shifting cultivators—again largely those farming in the Central Sierra. For example, a 1989 issue of the official magazine *Parques Nacionales* states:

> This historic process of deforestation, caused by the practice of shifting cultivation, indiscriminate use of forests for firewood and charcoal, the act of slash and burn…has produced grave consequences.…[T]o combat the increasing problems of deforestation and erosion, which are placing in danger the environmental and productive stability of the country, the present system of protected areas must be protected in the immediate future.

The Forestry Action Plan for the Dominican Republic (FAO 1991, 51) employs similar rhetoric:

> The extreme poverty of the inhabitants, associated to [*sic*] a low educational level and the lack of an adequate cultural heritage to make them aware of the proper management of natural resources, are the main causes of the ecological unbalance of Inoa, Amina and Bao Rivers watersheds.

A 1993 review of the state of the Dominican environment states:

> The principal obstacle that the Dominican Republic faces in its forest protection efforts is the large campesino population that inhabits mountainous zones and which most live by slashing and burning the forest cover as a principal survival strategy. (Martínez 1993, 89)

As late as 1997, Bolay writes:

> Doubtless this system (shifting cultivation) is a very dangerous one in the Dominican Republic of today, because it endangers the last remaining forests. Increasing population density reduces the fallow seasons,

so fertility is lost and erosion occurs. The rising urban population cannot be supplied by those production methods. (Bolay 1997, 138)

As del Rosario (1987) argues, this rhetoric overstated the role of small producers in environmental degradation and overlooked the very real contribution that traditional agriculture has made to food crop biodiversity and to Caribbean food security.

It is now less fashionable to blame environmental degradation on *conuco* agriculture. The failures of modernized agricultural systems throughout the island and a renewed emphasis on organic and fair trade production among consumers in the global north and in Dominican cities has had a perceptible effect on Dominican environmental rhetoric and on the culture of agriculture. National agricultural research centers are devoting more attention to the role and needs of the small cultivator. At the same time, responding to increasing consumer interest in organic and fair trade products on the part of Dominican consumers, the food industry is placing new emphasis on *conuco* crops (*comida tipica*) and organic production.

Before the revolution, small-scale food producers in Cuba suffered from the same stigma as their Dominican counterparts. The revolution saw a partial change in that food production was embraced as a national goal. However, revolutionary emphasis on adoption of technological methods resulted in a food production system and a devaluation of labor-intensive small farm food production strategies.[10] The Special Period saw a marked increase in government attention to food security and a tentative, but broad-based transition toward low-input agriculture, featuring integrated pest management programs, domestically produced organic fertilizers, vermiculture, and animal traction. Rapid expansion of food production on urban and rural land was a response to the failure of official channels (ACOPIO) to provide an adequate food supply for Cuban households, the opening of farmers' markets in all Cuban municipalities, the opening of *paladares* or small, private-sector restaurants, and the freedom of small cultivators on private lands to sell their surplus for dollars on the open market. The question is whether the flowering of alternative agriculture in the past decade masks a continuing emphasis on high-tech export production.

Embracing Agricultural Alternatives?

Can two small island nations, given their weak position in the global economy and their traditional denigration of local root crop production, hope to achieve domestic food security and agricultural sustainability? Given the substantial differences in their political systems, are Cuban and Dominican paths toward sustainability likely to differ substantially? If so, is one path likely to get there faster? In the following sections, I suggest that, while Cuba's national commitment to alternative agriculture is impressive, as is the scientific capacity it brings to this effort, in both nations the transition has been tenuous, partial, and fraught with problems.

Cuba: Low-Input Agriculture and the Special Period

Support for alternative agriculture may be stronger in Cuba than in any other low- or middle-income nation. The Cuban experiment with organic agriculture is a unique national effort to steer agriculture away from a dependence upon imported inputs without reducing the contribution of agriculture to export earnings. In the 1980s, Cuba depended on food and feed imports. The breakdown of trade relations with the former Soviet Union and Eastern Europe in 1990 resulted in about an 80 percent drop in pesticide and fertilizer inputs and about a 50 percent drop in the availablity of petroleum for agriculture. Spare parts for farm equipment manufactured in eastern Europe and cattle feed imported from the Soviet Union became scarce. Food availability declined in a country ideologically committed to domestic food security. According to Enriquez (1994, 1), its 1989 Food Program is "one of the—if not *the*—most important areas of government initiative in Cuba today."

Alternative Technologies. By 1990, the realization that Cuba could no long import agrochemicals, feed grains, or farm machinery on favorable terms led to policies promoting research on alternative technologies in agricultural research and training institutions. Thus, even before the limitations of the Green Revolution paradigm became apparent, Cuban interest in alternative approaches to farming systems was on the rise. In 1992, Castro embraced a strong environmental position at the United Nations Conference on Environment and Development (UNCED), further legitimating Cuban interest in alternative agriculture. Finally, in the early 1990s Cuban agricultural scientists came to view research on low-input alternatives as an area in which they enjoyed a comparative advantage. This was particularly true in the area of biotechnology (Díaz 1995; Díaz and Muñoz 1994).

The Cuban experiment is to a large extent technology—rather than farmer-driven. Rosset (1994, 2) found that "with only 2 percent of Latin America's population but 11 percent of its scientists and a well-developed research infrastructure, the government was able to call for 'knowledge-intensive' technological innovation to substitute for the now unavailable inputs." Alternative technologies have assumed a secure place in Cuban agricultural science with significant investments in research on biotechnology for pest control, innovative uses for sugar cane byproducts, vermiculture (the use of worms for soil enhancement), and meristem culture, a form of asexual crop plant reproduction.

Livestock production was hit hard by the loss of special trading relationships with eastern Europe. In the 1960s, revolutionary social policy emphasized milk and beef production, so that all Cuban citizens could enjoy a diet comparable to those of industrial nations. To achieve this goal, the state replaced locally adapted cattle with Holsteins and other improved, high-yielding breeds that required imported feed. As foreign exchange and feed imports became scarce, Cuban scientists devoted attention to sustainable forms of cattle production. In 1991, Cuban animal scientists introduced a rational pasture management system developed by France in the 1960s. Corn production for food and fodder expanded, and increased attention was paid to the development and rescue of native cattle breeds adapted to local fodders (Roberto García Trujillo, personal communication, 1993).

Finally, Cubans played a leading role internationally in improving animal traction. Responding to difficulties in obtaining farm machinery and oil, Cuba's sustainable agriculture strategy emphasized development of more efficient machines and the reintroduction of animal traction. Animals consume more energy, but they can be raised and fed locally, and they provide organic fertilizer. Rosset (1994) argues that oxen cause less soil erosion and do less damage to hillsides than tractors, but that they are far more labor-intensive. But even oxen are inappropriate on very steeply sloped lands, and they cannot be used where hills are stabilized by narrow terraces.

The Shift from Export to Food Production. In the 1990s, incorporation of land into the production of bananas, plantains, and root crops became a national priority, and, indeed, agricultural production is central to the Cuban conception of ecology.[11] One government initiative to increase food production was the construction of community, workplace, and urban gardens (Eckstein 1994; Rosset and Benjamin 1994; Cruz 1992, 1994). A second was diversification of production on existing agricultural lands (Levins 1993). About 22 percent of sugar lands were converted to food production in the 1990s (Enriquez 1994). A third was an effort to stem the flow of migrants out of traditional food crop–producing regions by improving the quality of life for rural communities.

Does the knowledge-intensive system described by Rosset and others represent a national shift to ecological agriculture, or is it simply a temporary import substitution economic strategy? The 2002 decision to curtail sugar production is a positive sign. Will the experiment be seen in the long run as a model for agricultural development in the global south, or will its practices be discarded once the special period is over and devalued as inventions of necessity? The long-run success of policies that have created new niches for low-input food production will depend upon changes in Cuban attitudes toward peasant production and toward typical Caribbean crops.

Changing the Culture of Agriculture: UBPCs and Farmers' Markets

In a 1991 speech to the Fifth Congress of the National System of Agricultural and Forestry Technicians, Fidel Castro argued that "we must convert farming into one of the most honored, promoted, and appreciated professions." And, indeed, Cuban cultivators have gained economic power in the past five years. With food scarcities, peasant agriculture took on new significance. According to Torres Vila and Pérez Rojas (1995, 4), "owing to the growing food scarcity, [campesinos] are subject to enormous pressures to become involved in private sales. As many have said, from the beginning of the Special Period, the number of visits from almost forgotten family members and fictive kin has grown substantially." By 1993, it was clear that the state sector would have to be re-formed to capture some of the energy present in the smaller private sector.

One organizational change was to allow state-farm workers to form Basic Units of Cooperative Production (UBPCs) on lands to be leased from the state rent-free for an indefinite period. UBPCs receive credit and services from the state, but they are expected to be self-sufficient. Responsibility for production lies with producers, but farmers have continuing access to the state research apparatus. They must sell a portion of their principal crop to the state, but they can sell the surplus as they see fit. In 1994, 40.6 percent of Cuba's agricultural land was in UBPCs, as opposed to less than 30 percent for state enterprises (Deere 1995). While some UBPCs experienced serious problems, others per-formed well. But the line between access and interference is fine, and some co-operative workers have viewed their UBPC as little more than a "state farm with a bank account" (Torres Vila and Pérez Rojas 1994, 12). Sustainability may also be enhanced by the requirement that UBPCs meet their basic food needs. This forces diversification in otherwise monocultural settings.

A second, smaller effort took place in the mountains of Escambray (Terrero 1996), where in 1994 the director of the state coffee enterprise made individual farm families eligible to receive usufruct rights to coffee plots in order to "try their luck" with individual farming. Again, farm families are expected to be self-sufficient and are likely to grow locally adapted root crop staples and raise live-stock, thus contributing to farming system reintegration.

The reopening of farmers' markets in 1994 allowed producers to obtain higher prices for their goods and to count on a higher demand (Torres Vila and Pérez Rojas 1995). Consumers have been highly receptive to the new markets, which permit the government to use taxes to channel the flow of goods to areas where the demand is greatest—notably Havana. Torres Vila and Pérez Rojas note that the markets encouraged growth in levels of production, greater diversifica-tion, and polyculture. The markets are not in and of themselves instrumental in the shift to alternative agriculture, but by allowing prices to fluctuate and elimi-nating the state as an intermediary, the new markets helped to create sustain-able livelihoods in agriculture for Cuban UBPC members and independent producers. In the 1990s, because their products fetched relatively high prices, peasant producers experienced a rapid rise in economic status (Sinclair and Th-ompson 2001; Enriquez 2003). These organizational changes alone are insuffi-cient to make Cuban agriculture sustainable, but they have contributed to an emerging culture of farming that values labor and knowledge-intensive mixed production along with hi-tech monoculture. In addition, by exposing cultivators to greater risks, they offer them the possibility of greater gain.

Urban Agriculture

Food production throughout the world is becoming an urban as well as a rural phenomenon, and in Cuba the amount of urban land under cultivation in-creased rapidly during the Special Period. From the top of the José Martí monument in Havana's Plaza de la Revolución one can see huge expanses of raised beds on what had once been unproductive lawns surrounding government buildings. Park and vacant lands are being put to agricultural use. Backyard gardens are encouraged, and, sometimes in the absence of official encouragement, livestock production has increased within city limits. Hydroponic and organoponic vegetable beds have been installed at factory sites and in urban neighborhoods (Cruz 1992). These gardens, which represent the formal end of the urban agricultural continuum in Havana, numbered 2,611 in 2001 (Peters 2001). Their produce goes to workplace cafeterias, local schools, and day care centers; surplus is often sold at small farm stands. Land preparation in Havana is hard work: elaborate stone walls testify to the labor required to turn the local karst into fertile ground. A major impetus for production has been the ability of individual cultivators to market their surplus directly. In this way, small-scale commercial producers earned in a month what government employees brought home in a year.

The Dominican Republic: Experiments at the Margins

Dominican experiments with alternative agriculture have occurred at the margin, carried out by NGOs, sometimes in collaboration with agricultural research institutions, at times with international financial support. But the general state of disarray within the Secretariat of State for Agriculture (SEA) and the National Institute for Hydraulic Resources (INDHRI) during the Balaguer regime prevented a major focus on urban food production, and most international investment in Dominican agriculture went to support high-input production of such nontraditional agro-exports as cut flowers and oriental vegetables. At a national level, before President Leonel Fernández took office in August 1996, no attention had been given to the withdrawal of land from agriculture or to the need to validate peasant food producers. Sustainable agriculture programs were focused on demonstration green marketing initiatives, conservation and agro-forestry projects on hill lands, and participatory research programs.

The Green Marketing Strategy
This approach to sustainable agriculture focuses on a marketable product whose culture is, in principle, associated with sustainable land use. A variety of programs promote cultivation of tree crops and perennials in agro-forestry or agro-silvo-pastoral systems. Oxfam UK has played a significant role in supporting organic coffee production in the Dominican Republic. In the early 1990s, the Asociación de Desarrollo de San José de Ocoa, with Ford Foundation assistance, promoted avocado cultivation on steeply sloped lands using microirrigation technologies. ENDA-Caribe successfully promoted production of trees and woody species in alley cropping systems (Rocheleau and Ross 1995).

International import standards have also stimulated change. Failure to meet USDA import requirements for pesticide residues had a serious impact on agro-exports. Murray (1994, 76) reports that in 1987–88, the Dominican Republic "held the dubious distinction of having the highest rate of illegal pesticide residues in samples of produce imported into the United States." Losses due to crop detention (principally in the oriental vegetable sector) were estimated at $2.5 million. Economic losses, coupled with rising concern among urban intellectuals about birth defect rates in the region around Constanza, stimulated interest in organic fruit and vegetable production.

Organic coffee production has been successful. Producer prices for organic coffee have been higher than those for sun-grown, non-organic beans. In the mid-1990s, few producers were able to market 100 percent of their crop at higher organic prices, but the organic market acted as a safety net for the small producer. However, attractiveness of organic coffee production over the long term will depend upon a steady demand for organic coffee in Europe and increasing demand in the United States.

Domestic outlets for organic production are harder to find, but several Dominican NGOs have helped to create urban outlets. The Comité para la Defensa de los Derechos Bariales (COPADEBA) established small stores in poor Santo Domingo neighborhoods that were intended to provide outlets for producer cooperatives. Grupo Ambiental Habitat has promoted sale of products from the Dominican Southwest, and ENDA Caribe has engaged in marketing studies for agro-forestry products. Notably, Dominican ice cream producer Helados Bon is buying organic coffee, cacao, and fruits for its premium ice cream. Still, many Dominican consumers lack the income elasticity or flexibility to allow environmental considerations to determine their choices. At the end of the day, the success of a green marketing strategy will depend upon either strong, stable local markets or a high degree of producer flexibility in response to market signals. In the absence of timely and accurate information, niche marketing in international or even regional markets is a risky peg on which to hang a sustainability strategy. Therefore, even greater efforts will be needed to expand domestic markets for organic and sustainable produced food.

The City on the Hill Strategy

NGOs and donors have found it tempting to create model programs for sustainable agriculture on lands considered marginal for use as pedagogical tools. The idea is that once farming systems based on agro-ecological principles are put in place, other farmers and extension personnel will adopt and further diffuse environmentally sound technologies to other areas.

The most visible agroecological experiment in the Dominican Republic has been Plan Sierra—a university experiment station that became a parastatal agency in the 1970s and an NGO in the 1980s. Plan Sierra often brought in international experts in agro-ecology and forest management to evaluate its programs and assist in their design; these experts lent enormous visibility to the project in development circles. Its efforts included sustainable timber management, pig breeding, and the development of alternative practices for steeply sloped lands. The experiment and demonstration station at Los Montones is an idyllic setting for short courses and farmer meetings. Its highly productive demonstration plot features such labor-intensive practices as terracing, herb cultivation, and alley cropping. In a region where employment opportunities outside of agriculture abound, there is some question as to whether such labor-intensive practices are replicable. Plan Sierra's city-on-the-hill status also made it vulnerable to criticism on other fronts as well. It set up producer organizations that paralleled and at times competed with existing organizations in the region. This organizational competition reduced the Plan's effectiveness. Also, Plan Sierra appeared paternalistic at times, and it was not always clear that its programs could be replicated in the absence of continuing infusions of government and international funds.

A second high-profile agro-ecological experiment was situated in the town of Río Limpio in the Haitian-Dominican Border Region. CREAR, a small NGO started by a former Peace Corps volunteer, introduced biodynamic agricultural practices into an erosion-prone region and sought to train cadres of "barefoot agronomists" to diffuse biodynamic agriculture to other similarly poor and ecologically vulnerable regions. CREAR staff developed farming systems that are highly productive over a sustained period of time, but their efforts at diffusion were less successful.

A third experiment has been the introduction of agro-forestry techniques by ENDA-Caribe in the Zambrano-Chacuey region. ENDA's major achievement was to broker an agreement with the Dominican forest service that exempted growers of tree crops from onerous regulations prohibiting any cutting of standing trees. While successful in many respects, the largely male NGO staff helped to put in place a structure that overlooked women's roles in agriculture and concentrated project benefits in men's hands (Rocheleau and Ross 1995).

"City-on-the-hill" sustainable agriculture projects have had two serious limitations. First, they require a lot of labor. Biodynamic and agro-ecological practices require careful timing and complementarity (Ploeg 1990). This makes them hard to replicate in a society where a growing portion of family income comes from off-farm work. Special incentives at a national level would be needed to encourage the long-term investments in labor that are crucial to these systems' viability. Second, management of city-on-the-hill projects has tended to be intensive and paternalistic. This can produce a backlash that limits their political and social sustainability.

Participatory Research Approaches

One way to address the problem of replication is to engage farmers in the development of sustainable practices. Rosset (1994) suggests that Cuba's MINAGRI has not gone as far down this road as some NGOs in developing countries. And, conversely, there have been some promising experiences in the Dominican Republic. However, in the Dominican Republic much work remains to be done in this area.

All three approaches hold some promise, but their impact remains marginal because they fail to address obstacles to sustainable agriculture at a national level—increasing inequity of land and income distribution, removal of productive land from agriculture, and degradation of prime agricultural lands due to poor water and pest management practices. Diffusion of sustainable practices on a national scale will depend on a reorientation of the SEA from export to domestic food production and from conventional to alternative practices. Unfortunately, over the years, SEA's efforts to work with small farmers have been minimal, and Dominican extension staff have lacked the resources and training needed to work with farmers on the development and diffusion of alternative practices.

Cuba and the Dominican Republic Compared

The major difference between sustainable agriculture programs in Cuba and in the Dominican Republic in the past decade has been the level of state support.[12] In Cuba, the state has fostered a commitment to alternative agriculture within the national research establishment. In addition, the unique role of alternative agriculture in Cuban agriculture became a source of both national pride and international prestige and legitimation for the Cuban state in the post-Soviet era (Guptill 1996; Koont 2004). In the Dominican Republic, in contrast, pressure for sustainable agriculture in the 1990s came largely from local and international NGOs, international donors, and foundations. Before 1996, state commitment to alternative agriculture was notably absent. This, however, appears to be changing, as organic products become more marketable than conventional products and as funding for NGO efforts becomes scarcer.

A second difference is the locus of sustainable agriculture programs. Dominican sustainable agriculture programs are concentrated in steeply sloped and marginal areas, dominated by smallholder production. In Cuba, sustainable agricultural practices have been introduced on a broader scale, within the UBPCs, cooperatives, community and workplace gardens, and individual holdings. Government encouragement of urban and peri-urban agriculture is almost completely absent in the Dominican Republic.[13] Finally, sustainable livestock management has received attention from the Cuban animal science establishment, while alternative livestock programs in the Dominican Republic have been restricted to small-scale experiments with household pig and small animal production.[14]

A third difference between the two countries lies in the definition of ecology. In the Dominican Republic, ecological systems are generally conceived as excluding human use. The result has been a strong bifurcation between natural areas and agricultural land, which has led to policies closing large areas of the country to agricultural production in any form. Since 1992, Cuban use of the term "ecology" has implied human activity. For example, strategies for improving the ecology of Havana's huge Parque Metropolitano incorporate and encourage agricultural land use for domestic food production.

Perhaps the most important factor influencing adoption of alternative agricultural practices has been the global economy. Cuba faced a severe and sudden food crisis in the 1990s owing to its loss of favorable trading relationships with eastern Europe; in contrast, the Dominican Republic experienced a gradual deterioration in terms of trade. So, in the 1990s, the impetus for the search for alternative food production strategies was far stronger in Cuba than in the Dominican Republic. To some extent, Cuban emphasis on agricultural sustainability and food security may have been an artifact of the Special Period. Cuban borrowing in the 1990s to purchase conventional inputs for sugar and tobacco production suggested that the emphasis on low-input, sustainable practices would remain confined to domestic food production. But even Cuban planners and social scientists working in the latter sector in the late 1990s were guarded in their optimism, concerned that if producers could regain access to agro-chemicals, they would be likely to use them. It may be that transformation of the culture of agriculture has been sufficient to make these concerns unwarranted. Sinclair and Thompson (2001) find a divergence of opinion within the Ministry of Agriculture, but they suggest that farmers who combine organic and conventional practices will be highly responsive to market signals.

The transition from conventional to more sustainable farming practices demands creation of alternative marketing systems that can guarantee green producers a modicum of price stability. There has been a great deal of talk about green marketing, but little research. Recent research on marketing initiatives in Cuba (Torres Vila and Pérez Rojas 1995) suggests that government incentives can help to match supply and demand without creating cumbersome bulking and retailing infrastructure. Dominican NGOs have been seeking to develop urban outlets for the products of sustainable agriculture. But there is a need for further research on the economic viability and vulnerability of green and niche marketing.

Prospects for Change

To summarize, development of sound, locally appropriate land and water management practices is a good indicator of a commitment to achieving agricultural sustainability. Cubans have made considerable progress on this front with their emphasis on biofertilizers, biological crop protection, improved water management techniques, animal traction, and improved pasture management. Its strong scientific establishment and continuing research emphasis on ecological agriculture have positioned Cuba to play a leading role in this effort. However, no research establishment can carry out development and monitoring activities across the spectrum of microclimatic and ecological zones. Rapid dissemination of information about alternative practices must be supplemented by on-farm monitoring and evaluation. This entails active involvement of cultivators in the quest for sustainability. Second, adoption of alternative technologies does not guarantee sustainable agriculture. If both countries' economies remain dependent on imports, pressures to divert resources for food production into exports will continue. Where agribusiness standards and practices favor conventional agriculture, export agriculture will not be sustainable. And, as the history of sugar shows, even when faced with negative price incentives, both Cuba and the Dominican Republic have had to overcome considerable inertia to abandon conventional agriculture—especially where there are massive sunk costs in equipment and infrastructure. Conversion to meet international organic standards also entails substantial investments, and these are likely only where markets for organic products are strong.

Cuban and Dominican alternative agriculture efforts have also suffered from two liabilities that have limited their scope and viability. One was an emphasis on marginal lands—more pronounced in the Dominican case. A second has been a tendancy to promote "magic bullet" solutions whether or not they are ecologically or economically appropriate. Diffusion of inappropriate technologies can make farmers wary of new programs emphasizing low-input practices, and even successful practices may have negative consequences for women or weaken the bonds that tie households and communities together (Schroeder 1993; Rocheleau and Ross 1995). Finally, even where new introductions have proven ecologically, economically, and socially successful, more data on yields and environmental performance are needed if these practices are to find acceptance within agricultural research establishments.

The strength of these braking forces makes preservation of the agricultural land base and native plant and livestock varieties even more important. But preservation has to be seen as a step forward, rather than an attempt to repair the fabric of colonial agrarian societies. In urban societies, this will require a cultural effort to foster appreciation for agriculture as an urban phenomenon. In 1991, when Fidel Castro, addressing the Fifth Congress of the National System of Agricultural and Forestry Technicians, stated, "We must convert farming into one of the most honored, promoted, and appreciated professions" (Rosset and Benjamin 1994), he acknowledged the cultural dimension of agricultural sustainability and the importance of building a culture in support of agriculture within an urban context.

Sustainable agriculture movements can build on urbanites' positive associations with the garden and with rural landscapes. We can look to rural land preservation movements in North America and Europe for ideas and examples, but with the realization that these models are not transferable. There may be lessons to be learned from western Europe and Canada about linking tourism to agricultural land preservation. Increasing interest in the islands on the part of Caribbean populations in the United States could be channeled into agritourism efforts. Agritourism now takes place on an informal basis in the Dominican Republic as urban dwellers and Dominicayorks return in search of their rural roots. It could certainly be encouraged in Cuba.

The transition to sustainable agriculture in both these countries is underway, but it is still hard to tell whether experiments in either country can be sustained over time or diffused widely throughout the agricultural sector. Earnings from export agriculture are dropping, but as long as conventionally produced crops remain important sources of foreign exchange the potential to mainstream alternative agriculture will remain limited. On the other hand, agricultural sustainability and food security in the Caribbean are likely to become more and more closely tied in the years to come. The special period may turn out to have been not so very special after all.

Notes

1. The first component has received increasing attention in the past decade. Development of agro-forestry systems, erosion control, and integrated nutrient and pest management practices are all indicators of progress on this front. The broad outlines of alternative farming systems are laid out in Altieri (1987), Altieri and Hecht (1990); and Carroll, Vandermeer, and Rosset (1990). As a result of this early emphasis on the farm field, sustainable agriculture discussions often focus on development of on-farm land, water, and pest management practices that minimize soil erosion, maintain fertility, and reduce dependence on chemicals for pest control. Hillside agriculture and shifting cultivation have received a disproprotionate share of attention in sustainable agriculture debates.

2. See Murray (1994) for a discussion of the impact of white fly eradication programs on tomato production in Azua, Dominican Republic.

3. The experience of the Dominican Republic is somewhat different in that, for the most part, its agro-export sector was not a legacy of Spanish colonialism, but rather developed toward the end of the nineteenth century with the introduction of industrial sugar production. Until the 1870s, Dominican rural society was characterized by livestock production, communal forms of land tenure, and informal food production systems. For excellent studies of Dominican rural society in the late nineteenth and early twentieth centuries, see Hoetink (1982), Baud (1995), Franks (1997), Turits (2003), and San Miguel (1997).

4. Thrupp and Pérez (1989) note that conservative Cuban agricultural researchers and officials in the Ministry of Agriculture (MINAGRI) allied with pesticide manufacturers and importers in opposition to biological control strategies. For detailed discussion of the link between Green Revolution agriculture and the petrochemical industry in Mexico, see Wright (1990).

5. As compared to 10.4 percent for Cuba, 12.7 percent for Costa Rica, and 13.3 percent for the United States.

6. Estimates range from 10.4 percent for 1994 (World Resources Institute 1996) to 12 percent for 1995 (Ministry of Science, Technology, and Environment 1996, 6).

7. This is not to suggest that the state plays no role in shaping agricultural strategies. Emphasis on exports at the expense of food production may reflect states' interests in enhancing their own autonomy and in creating conditions for their survival in a hostile political and economic environment.

8. According to an *Economist* (1996) survey, in 1990, the sugar harvest was a record 8.1 metric tons, and it fell to 3.3 metric tons in 1995, the lowest in fifty years. Heavy borrowing at high interest rates (14 percent) made possible the purchase of fuel, pesticides, and spare parts for aging farm equipment.

9. Efforts to encourage individual cultivators to pool their holdings in cooperatives

enjoyed varying degrees of success, although crops like coffee and tobacco have remained resistant to collectivization (Deere, Meurs, and Pérez 1992; Torres Vila and Pérez Rojas 1994). And, at present, the Cuban government is encouraging individual coffee producers (Terrero 1996).

10. Rosset compares the evolution of Cuban agriculture to that which took place in California's Central Valley (Rosset and Benjamin 1994).

11. See, for example, Cuba's Agenda 21 report (Ministry of Science, Technology and Environment 1995).

12. Sinclair and Thompson (2001, 18), outlining agricultural development policies and practices in Cuba and the Dominican Republic, offer interesting comparative data on land tenure, farm income, caloric intake, and the role of agriculture in national economies. Their very useful table does not, however, capture either the contraction of export agriculture in the Dominican Republic or the movement of export agriculture onto small farms with the introduction of contract agriculture in the 1980s.

13. In Santo Domingo working-class neighborhoods, the combined forces of commercialization, in-filling of urban neighborhoods (Pérez 1996), and a building boom have all but eliminated spaces once dedicated to urban agriculture. In Santiago, the building boom of the 1990s created a new pattern of shifting cultivation, as land speculators and builders encouraged guardians to cultivate their vacant holdings within the city that were being reserved for high-rise buildings (del Rosario 1987; Yang 2000) in order to prevent informal settlement or solid waste dumping. This new migratory agriculture (to use del Rosario's descriptor) is a precarious phenomenon. By 2003, the number of plots inside Santiago's urban core had markedly diminished, but their number may increase as the impacts of economic recession are more widely felt.

14. For a discussion of USAID's disastrous program to eliminate swine fever by exterminating native pigs and replacing them with U.S. "improved varieties" dependent on high-quality feed, see Bolay (1997, 281–84). Plan Sierra and other programs directed at small farmers sought to undo the damage by interbreeding the U.S. varieties with feral pigs who had escaped the slaughter.

CHAPTER 8

"Ni Una Bomba Mas"

REFRAMING THE VIEQUES STRUGGLE

KATHERINE T. MCCAFFREY AND SHERRIE L. BAVER

Introduction

For decades, residents of Vieques, Puerto Rico fought a David and Goliath battle against the U.S. Navy. Until 1999, however, few people in the United States had ever heard of Vieques and its problems. Vieques is a 51-square-mile island, roughly twice the size of Manhattan, where more than nine thousand people lived wedged between an ammunition depot and a live bombing range. Since the 1940s, when the Navy expropriated more than two thirds of the island, residents have struggled to make a life amid the thundering of bombs and the rumbling of weapons fire. The U.S. Navy contended that the Vieques installation played a crucial role in naval training and national defense. The civilian community of Vieques argued that the military control of land and live-fire exercises caused severe ecological destruction, cancer and other health problems, and overwhelming social and economic crises.

A grassroots movement against naval operations emerged in 1978 led by local fisherman whose livelihoods were disrupted by the naval training exercises. Although similar protests in the early 1970s forced the military out of nearby Culebra, the Vieques struggle died out. With the end of the Cold War, Viequenses began to organize again. However, the struggle became widely known only in the spring of 1999 when the death of a civilian security guard sparked a new wave of protest and placed Vieques on the international stage.

The aim of this article is to understand how a local struggle became a national and international *cause célèbre*. Our hypothesis is that reframing opposition to military training in terms of environment, health, and human rights concerns allowed a broad coalition to form that reached well beyond party lines both in Puerto Rico and on the mainland. This coalition built overwhelming support for an end to live bombing exercises on Vieques, the Navy's immediate departure, and the return of federally controlled land to local authorities. In May 2003, activists achieved a major victory with the exit of the Navy from their community. The struggle, however, is now in its second stage: seeking social and environmental justice for Vieques.

Theoretical Overview

Conflict in Vieques had its foundation in the material conditions of everyday life. Vieques is the poorest municipality in all of Puerto Rico, with 73 percent of the population living below the poverty line. It has among the highest rates of unemployment—almost half the adult population is without work (U.S. Census Bureau 2000). It has among the highest infant mortality rates, and a growing rate of cancer and other health problems that residents believe have been caused by weapons testing.[1] This socioeconomic and health crisis connected to the military presence on the island and has been at the heart of civilian grievances against the Navy.

Nonetheless, Vieques's struggle to evict the Navy was consistently mired in complex political issues that sidelined community concerns. The Navy's use of the island for live-fire exercises invariably raised the divisive issue of Puerto Rico's political status as a non-sovereign U.S. territory. While a majority of Puerto Ricans support continued political and economic association with the United States, they also maintain a strong sense of a separate Puerto Rican national identity (Barreto 2002; Dávila 1997; Duany 2000; Morris 1995). Live bombing exercises on this inhabited island suggested the second-class status of Puerto Rican citizenship and inspired charged debate over national identity, patriotism, and loyalty. Activists from Vieques and Puerto Rico who were concerned specifically about military actions became embroiled in debates over sovereignty. The U.S. Navy interpreted objections to military operations, however specific, as ideologically motivated and a threat to national defense. The Commonwealth government tried to avoid confrontations that could jeopardize its relationship with the United States, particularly as they affected ongoing debates about potential statehood or a modified form of association.

With the Cold War's end and before the beginning of the War on Terror, the political terrain shifted. Grievances against the Federal government, and the U.S. military specifically, could be aired without the aggrieved simply being tarred as "radical *independentistas*." Still, protestors had to pick their battles and their tactics with care to get a wide airing of their issues. By the mid-1990s, activists in Vieques framed their grievances in a way that allowed for broad, nonpartisan support for a long-simmering just cause. Protest against the military, long embroiled in stifling debates over Puerto Rico's political status, became part of a nonpartisan movement for social justice. Humberto García-Muñiz, for example, suggests that activists' use of a human rights discourse allowed the Vieques protest to gain national and international allies (García-Muñiz 2001). Leading Protestant and Catholic clergy in Puerto Rico espoused Vieques's cause as a valiant struggle for peace and human rights.[2] In this chapter, we explore how a focus on environmental and health concerns also played a key role in this political shift.

The argument set out here builds on the work of others who have studied the social construction of issue framing.[3] Most relevant here is Margaret Keck's work on successful protests of the Acre Rubber Tappers' Movement in Brazil. Keck found that it is not necessarily easy to link social and environmental agendas, but those who can do it successfully combine "strategic acts of image making, alliance building, and the seizing of institutional opportunities."[4] Specifically, it is useful to have a good story that blends both environmental and social justice components; successful "causal stories" involve one set of persons doing harm to another set. The joining of stories widens not only the potential pool of participants, but also the pool of public who will respond to one or another element in the story. While the rubber tappers in Acre and the residents of Vieques had reasonable environmental and equity complaints, Chico Mendes's murder in Brazil and David Sanes's accidental death in Vieques gave these stories a poignancy and urgency they might otherwise have lacked. In sum, reframing longstanding struggles for social justice in terms of environmental concerns allows disadvantaged people to expand their resources and strategies considerably.

The Navy's Arrival on Vieques

The U.S. Navy expropriated more than three quarters of Vieques Island in the 1940s. The military was impelled to take Vieques land first by the perceived German threat in the Caribbean during World War II, and later by mounting cold war tensions.

In 1941, declaring a national emergency, the Navy seized 6,680 acres of land in Ensenada Honda, on Puerto Rico's east coast, and 21,020 acres on Vieques, two-thirds of that island's land, to build Roosevelt Roads Naval Station, its most important operating base in the Caribbean. Roosevelt Roads was planned to rival Pearl Harbor in scale and significance. The base would stretch across the Vieques Sound to connect Ensenada Honda and Vieques. Roosevelt Roads would provide anchorage, docking, repair facilities, fuel, and supply sources for 60 percent of the U.S. Atlantic fleet. Furthermore, with the threat of a German invasion of Great Britain looming, naval planners saw the base as a potential point of supply, repair, and refuge for the entire British fleet. The plan was to excavate tons of rock from Vieques in order to build a sea wall between the island and a new homeport in Ensenada Honda. Huge magazines would be cut into the hills of Vieques, and a Marine camp established on the neighboring island of Culebra (Langley 1985, 271–75; Tugwell 1977, 68).

Vieques's stark social inequality and overwhelming poverty facilitated the military takeover. Vieques's economy was dominated by sugar cane monoculture. Ninety-five percent of the rural population, or two-thirds of the total population of 10,582, was landless, while two sugar corporations occupied 71 percent of island land.[5] The concentration of the land in the hands of two sugar corporations and a few wealthy farming families eased the transfer of two-thirds of the island from private to military control. The landless majority that lived on sugar land had little political clout with which to counter the U.S. Navy, and they were summarily removed from their homes.[6]

Any economic gain from the military presence proved elusive. While base construction initially created an economic boom, the work stopped almost as quickly as it had begun. Military priorities shifted to the Pacific when the United States entered the war, and the Navy scaled back its original plans. Construction of the breakwater from Puerto Rico to Vieques was suspended, and work on Roosevelt Roads slowed because of a shortage of supplies. By the time Roosevelt Roads was completed in 1943, naval planners had concluded that a major naval base in Puerto Rico was unnecessary. Roosevelt Roads was placed on maintenance status at the conclusion of World War II (Langley 1985, 272–73).

The abrupt halt of construction had a devastating effect on Vieques's economy. Without the military project, there was no work left on the island. The Navy's expropriations of land had effectively liquidated the sugar cane industry. Playa Grande's *central*, the last operating mill in Vieques, was dismantled and sold. Most sugar cane lands had become military property, part of either the base itself or the resettlement tracts where tenants were relocated. Though some small- and medium-sized independent farms remained, the farmers had no mill to which they could sell their harvest, and no access to transportation to ship their cane to the mainland to be ground. Residents were left languishing in squalid resettlement tracts without a clear sense of the future. They were assigned plots without title to the land and were not allowed to transfer lots. Residents were warned that they would be evicted if the Navy wanted to reclaim the land (U.S. House of Representatives 1981, 3).

The lack of title to the land had a number of damaging effects. It was impossible to secure loans to build decent homes. The lack of property rights left unanswered issues of inheritance, raising questions as to whether an individual's child would hold any rights to the house or land where he or she was raised. It is possible that the Navy saw residents as a valuable labor force during the frantic construction of the pier. But circumstances changed in 1947.

In 1947 the Navy drew up new strategic plans for Vieques. The Navy redesignated Roosevelt Roads as a Naval Operating Base for use as a training installation and fuel depot (Langley 1985, 273). Vieques would be converted into a training site, to be used for firing and amphibious landing practice by tens of thousands of sailors and marines. Because this new vision of the base required more land, the Navy planned to expropriate more than four thousand acres from eastern Vieques, displacing 130 families (*El Mundo*, 6 June 1947).

The second round of expropriations would wedge Vieques's population between an ammunition depot and a maneuver area. The Department of the Interior and the Insular government conducted closed-door meetings to address alternatives to squeezing the civilian residential community between the two installations. The Department of the Interior proposed resettling the population in St. Croix, just as the U.S. military had usurped Bikini Island the year before and deposited its inhabitants on a distant atoll.[7] The Puerto Rican government managed to block the eviction of Viequenses from the island, but the naval expropriations prevailed. Wary of the spectacle of dislocated families dumped on the street, the Puerto Rican government agreed to build housing in a new resettlement tract. The only concession the island government won from the Navy was that the military would provide materials for this new construction. Another minor provision permitted continued cattle grazing on the western area of the military property used for ammunition storage. Since naval maneuvers would include launching live bombs, the rest of the island was too dangerous for civilian entry. The Navy justified its plans by pointing to a "changed international situation"—namely, the perceived threat of communist proliferation across the world (*El Mundo*, 16 October 1947).

With three-quarters of its land usurped, Vieques's quality of life was severely debased and its economy crippled. Residents were sandwiched between an ammunition depot and a vast maneuver area. The municipality lost its tax base. The inhabitants of this arid tropical island lost access to major aquifers on military-controlled land. Although only six miles lay between Puerto Rico and Vieques, ferries were now forced to travel a circuitous twenty-two-mile route in rough waters to avoid the Navy danger zone. The military dismissed islanders' worries, arguing that the economic benefit of spending by the troops would offset any economic problems. This prediction did not come to pass.

Though the Navy argued that the military installations in Vieques would provide work and opportunity to islanders, they brought instead ongoing unemployment and poverty. The number of troops permanently stationed on the island was not large enough to promote the development of a service economy. Vieques was used primarily for ammunition storage and maneuvers, and secondarily as a Marine base (1959–1978). Thousands of troops would pour onto the island in the 1950s and 1960s, but their visits were so sporadic and brief that they could not sustain the local economy. In the early 1960s, the Navy drafted secret plans to evict the entire civilian population and take over the island. An executive order from President Kennedy prevented the forced removal of residents, but underlying antagonism remained.

The History of the Vieques Struggle against the Navy

Stage I: The Struggle of the 1970s

In the late 1970s, the Navy intensified maneuvers on Vieques Island after a militant anti-colonial movement evicted the military from the neighboring Puerto Rican island of Culebra. In order to understand the protest movement that unfolded in Vieques, it is crucial to examine briefly the Culebra struggle and the way it defined protest against the military as anti-colonial in character. Culebra and Vieques formed a strategic triangle with the Roosevelt Roads

Naval Station on the main island. The Navy launched amphibious assaults on Vieques and concentrated naval and aerial bombardments on Culebra. Heightened bombing and naval efforts in 1970 to evict Culebra residents to expand the bombing range sparked a protest movement there.

The Culebra movement erupted during a charged moment in Puerto Rican history. A wave of resentment against Puerto Rican conscription into the Vietnam War catalyzed the independence movement. Student protest against the draft and the war in Vietnam gripped the island. There were mass demonstrations on the streets of San Juan against the "slavery of the draft." Clashes between *independentista* students and ROTC cadets at the University of Puerto Rico escalated into bloody riots. The naval presence in Culebra came to signify to the Puerto Rican Left the very essence of Puerto Rico's colonial subjugation to the United States. While grievances were material, the battle of Culebra became defined in terms of Puerto Rican independence. The Puerto Rican Independence Party and the Puerto Rican Socialist Party led a spirited direct action campaign to evict the Navy from Culebra.[8]

The Culebra movement succeeded in removing the Navy from that island, but the military ultimately prevailed. The Navy simply shifted its bombing practices from Culebra to Vieques. The movement that erupted in Vieques was a direct response to heightened maneuvers after the Navy pulled out of Culebra. Vieques residents objected to the naval presence because of a discrete set of material grievances. Protesting the Navy presence, however, was widely perceived as communist-inspired and anti-colonial in character.

Fishermen emerged as key protagonists in Vieques's struggle. The Navy's intensified maneuvers created particular hardships for the island's fishermen. Bombing caused great damage to coral reefs and fish populations in an already fragile marine environment. As ship traffic increased, Navy boats frequently severed buoy lines from the traps they marked, effectively destroying fishing gear and the financial investment the traps represented. Fishermen were prevented from entering waters they claimed were the best fishing grounds around the island. A newly organized fishing cooperative became a crucible of the anti-Navy movement. Fishermen launched a direct action campaign, interrupting international military maneuvers in ten-foot-long wooden fishing boats.

Fishermen created unity where there were divisions. Most people in Vieques were reluctant to confront the military since protest was commonly construed as anti-American. The valiant struggle of the fishermen, however, emphasized the economic nature of local grievances. It drew on evocative cultural imagery of Puerto Rican rural traditions in conflict with modern warships and weaponry. The fishermen's struggle became the basis of a coalition movement. A group called the "Crusade to Rescue Vieques" formed to back and expand the fishermen's war into a sustained movement to remove the Navy and reclaim the land. *Independentistas* in Puerto Rico and the United States found a new cause for their struggle.

The mobilization lasted for approximately five years. Vieques support groups sprang up throughout Puerto Rico and the United States. Congress held hearings on the status of naval activities in Vieques. Yet unlike activists in Culebra, Vieques demonstrators were unable to evict the Navy. By 1979 the political climate had changed markedly. The Soviet Union invaded Afghanistan. A wave of revolutionary movements swept Central America and the Caribbean Basin, heightening Washington's anxiety about the spread of communism and the growing influence of Cuba throughout the region. The Navy was determined to expand its presence in the Caribbean and maintain its Vieques installation.

Navy officials dealt with protestors with a heavy hand, arresting demonstrators who entered military land on trespassing charges and pushing Vieques's struggle into federal courts that were overtly sympathetic to the military. Protest came to be treated as a threat to national security. One protestor died under suspicious circumstances while serving a six-month sentence in federal prison on trespassing charges.[9] The Navy also launched a public relations blitz, seeking to discredit the movement as communist-inspired and led by outside agitators. While fishermen had oriented the movement toward economic grievances, the Navy undermined unity by refocusing debate on issues of patriotism and political affiliation. Tensions erupted between fishermen, local activists, and *independentista* supporters in Puerto Rico over the leadership and character of the movement.

When Puerto Rican Governor Carlos Romero Barceló intervened in the conflict in 1983 to sign a memorandum of understanding with the Navy, he effectively defused the movement. Romero essentially extracted a "good neighbor" agreement from the military, in which the Navy agreed to bring jobs to the island and work to mitigate the negative environmental impact. Leading activists were furious, swearing that the governor had handed Vieques over to the Navy on a silver platter.

Yet after years of bitter struggle and divisiveness, many Viequenses embraced the accord as the resolution to a long, difficult conflict. By bringing the Navy to the bargaining table, the accord seemed to acknowledge the legitimacy of local claims and offer at least a symbolic victory. The Navy, which had long dismissed all local claims as unfounded and which had characterized the movement as a communist insurgency, now admitted that it had done wrong. The Navy promised to help the local economy and to make efforts to mitigate its damage to the environment. It recognized its obligations to be a good neighbor and to strive to improve the welfare of island residents. More than any other event, the accord contributed to bringing about an end to the local movement and the decline of political activism in Vieques.

Stage II: Rebuilding a Movement

By the 1990s, it was clear that the Romero accord was a failure. Though the Navy promised to bring full employment, unemployment rates had risen to levels higher than when the military signed the agreement. The Navy's environmental commitments seemed more rhetorical than real, aimed mainly at improving the military's public image. While it brought an end to organized protest, the Romero accord did not dispel the core resentment that had motivated that movement. The Navy maintained control over the majority of island land and resources, and therefore over the future of Vieques.

Activists took advantage of a changed political climate in the early 1990s to try to rebuild a movement to reclaim land. The time seemed propitious. The Cold War was over and the Clinton administration announced formation of a Federal Commission on Base Closures, headed by Defense Secretary Les Aspin. In 1993, activists founded a group called "The Committee to Rescue and Develop Vieques" (El Comité pro-Rescate y Desarrollo de Vieques/CRDV) that aimed to include Vieques in discussions of which bases to shut down.

The mobilization that began in 1993 differed from the one in the 1970s. Although the core of the new committee was *independentista* and leftist in political orientation, the 1993 activists strove to build bridges to more moderate, centrist constituencies. In addition, the group changed tactics, moving away from direct confrontations and adopting more mainstream political strategies such as lobbying public officials and searching for compromise. Finally, the 1993 Committee distinguished itself from the crusade of the 1970s by concretely focusing on Vieques's future development without the military. Puerto Rican and mainland planners were consulted to develop a blueprint for land use in which the majority of Viequenses, rather than wealthy, off-island developers or local speculators, would enjoy the fruits of development.

The moderate rhetoric and mainstream tactics paid off in support for the Committee from the municipal government of Vieques and the Puerto Rican legislature. Furthermore, the Committee's work caught the attention of Carlos Romero Barceló, now Puerto Rico's Resident Commissioner in Washington. Well aware of the failure of his 1983 agreement with the Navy, Romero attempted another compromise with Washington. In 1993, Romero submitted the "Vieques Land Transfer Act" to the U.S. House of Representatives, proposing the return of roughly 8,000 acres in western Vieques to the *municipio* for public purposes. The Navy would continue to use land on the eastern side of the island for weapons testing and maneuvers. This proposal was reasonable, since the land on the western side of Vieques was, in reality, a vast, empty tract checkered with more than a hundred ammunition magazines. According to the Navy, forty of these bunkers were inactive. This meant that sixty bunkers were monopolizing 8,000 acres—nearly one-third of Vieques Island. Romero's bill had the potential to act as a pressure valve, allowing the Navy to keep much of its holdings while appeasing local citizens.

The Navy, however, firmly opposed the bill, and its response in the spring of 1994 showed an utter disregard for local concerns. The end of the Cold War, argued the military, had only ushered in a new era of violent peace. In this new world order, Vieques would remain a vital training ground. The base would be essential to train for missions in the new Latin American drug wars, as well as interventions in the Caribbean, the Persian Gulf, and the Balkans. With the support of the pro-statehood Rosselló administration in San Juan, the Navy confidently announced plans to build a nine-million-dollar "Relocatable-Over-the-Horizon-Radar" (ROTHR) installation on Vieques. The ROTHR technology was originally developed during the Cold War to monitor Soviet fleets in the Pacific northwest. Now the sophisticated system would have a new purpose—to scan the Caribbean and Latin America for aircraft carrying illegal drugs to the United States.

The installation would consist of three parts: a transmitter located on Vieques, a receiver in Lajas, Puerto Rico, and an Operation Control Center in Norfolk, Virginia. The Vieques transmitter would include thirty-four vertical towers ranging in height from 71 to 125 feet; constructing this facility would require approximately one hundred acres of leveled land. Vieques activists were furious. Though the project was described as part of the War on Drugs, Committee members felt such a claim was a subterfuge for entrenching the military presence. After all, they reasoned, it was only in the midst of new efforts to recover western land that the Navy suddenly found new use for the land, which had lain idle for decades. The radar installation energized the Committee to offer a forceful response. For the first time in the long struggle against the military presence, activists framed their response with a focus on health: they would alert the public to the potential health dangers of the electromagnetic radiation the radar installation would emit.

Stage III: No Radar!

Anti-Navy activists tapped into the growing concern on Vieques regarding the health effects of the naval presence; such concerns dated back at least five years, to an article published in a Puerto Rican engineering journal about high concentrations of explosives in local drinking water (Cruz 1988). For more than five years, then, concern had been increasing on Vieques not only about contamination from military explosives, but also about reports of high levels of certain types of cancer in the community. The secretive nature of military activity and Viequenses' lack of access to information understandably intensified fear and suspicion of the Navy.

Soon after the Navy's announcement of its radar project, public hearings took place in Vieques City Hall. The Committee organized a demonstration outside the hearings to boycott the meeting, while inside, a handful of anxious local residents raised critical concerns including the aesthetic appearance of the facility on land that locals had hoped to conserve for ecotourism and the health effects of electromagnetic radiation. The local newspaper described the opposition to the project as composed of "sympathizers from all parties, various religious groups, municipal assemblypersons and even government employees, many not considered anti-Navy types" (*The Vieques Times* 1994). Clearly, the Committee had successfully surmounted locals' fears of communist labels and had organized a demonstration that drew a diverse group of community residents. After weeks of reflection, the majority of Committee members decided that emphasizing the public health and environmental consequences was the best way to defeat Goliath.

As a sign of the successful reframing of this issue, a new group joined the local coalition—the Vieques Conservation and Historic Trust (VCHT). The VCHT was the pet project of some wealthy North American seasonal residents who were concerned, in particular, about preserving Vieques's bioluminescent bay. In the past, the Vieques Conservation and Historic Trust not only had refused to speak out against the Navy but were strongly supportive of the military presence. Therefore the Trust's statement of unequivocal opposition to the radar station represented a remarkable change of heart.

Over the next year, Vieques activists merged with activists in Lajas, Puerto Rico, to fight against the Navy's plans. In Lajas, radar opponents stressed the theme of agricultural land usurpation, while the Viequenses continued to emphasize health concerns. The crucial symbolic link between the two mobilizations was a group of Puerto Rican Vietnam veterans who focused attention on the issue of military contamination. In particular, one decorated veteran suffering from Agent Orange exposure served to undermine the moral credibility of the Navy's claims about the safety of the proposed ROTHR installation.

The Navy's continued refusal to compromise, coupled with several public relations failures, led to the October 1995 protest in San Juan against the radar project. This event represented one of the largest mobilizations in recent Puerto Rican history. The struggle over ROTHR continued over the next two years, including a brief moratorium on the project when the island's Environmental Quality Board (EQB) demanded detailed information about the health and environmental implications of the facility. The Navy persisted, however, and in 1996 announced its intentions to go forward with the radar project. The response in Vieques was a well-attended local protest in February 1997. Importantly, in this "Walk for the Health of Vieques," those organizing the protest were careful in the framing of their rhetoric. This mobilization was not to be seen as an anticolonial action but rather an event to dramatize the community's concern about perceived high cancer rates and other illnesses stemming from existing environmental contamination, as well as about future risks from electromagnetic radiation. Notwithstanding popular opposition, the Navy erected its ROTHR project in 1998. Ironically, to erect the antennae on Vieques, the military razed one hundred acres of mahogany trees that it had once claimed as one of its own major contributions to the ecological rehabilitation of the island.

The most important legacy of the 1995–1997 struggle over the radar installation was that Viequenses became organized and forged a wide coalition by focusing on health and safety concerns. This experience laid the foundation for the dramatic mobilization that erupted in April 1999.

Stage IV: The Death of David Sanes

On April 19, 1999, David Sanes, a civilian security guard employed by the Navy, was patrolling the Vieques live impact range. At one point during

Sanes's shift, two F-18 jets involved in training exercises dropped their two 500-pound bombs, but they missed their mark by a mile and a half. The Navy's range control officer and three security guards inside the training observation post were injured. Sanes, standing outside the observation post, was knocked unconscious by the explosion of shattered glass and concrete and bled to death from his injuries.

The Sanes family, while wanting no part in politicizing David's death, agreed to enter military land with Vieques activists to erect a large white cross in his memory. After the ceremony, which included christening the spot "Monte David," a well-known, self-proclaimed "environmental warrior" from Vega Baja, Albert de Jesus (a.k.a. "Tito Kayak") stole the spotlight. He personally pledged to camp at the site and block the resumption of military maneuvers. Over the next year, thousands of supporters from Vieques, Puerto Rico, and the U.S. mainland set up camps on the target zone and brought further military maneuvers to a halt. In this way, the encampments moved the Vieques struggle out of the local arena and into the national and international political spotlight.

The success of this mobilization is in large measure a result of the ability of the Committee to Rescue and Develop Vieques to create a resonant cultural framework throughout Puerto Rico.[10] Still another piece of the explanation is a continued focus on environmental and health concerns as the basis for the anti-Navy discourse. What was new in this stage of struggle was that the use of cultural and environmental/health frames drew support from large numbers of Puerto Ricans, regardless of partisan loyalties. This is a highly unusual state of affairs in Puerto Rico's political arena.[11] Also new to this stage was the large number of female participants—not only progressive feminists but also formerly apolitical women concerned about their community's well-being.[12] A third new element in this stage of the struggle was the active participation of major Puerto Rican religious institutions, who insisted on nonviolence as a tactic. All participants in the mobilization starting in 1999 were in firm agreement that no meaningful buffer zone could exist on a small, inhabited island like Vieques when the military tests weapons there.

The depth of support for Viequenses in Puerto Rico forced Governor Rosselló in May 1999 to form a Special Commission on Vieques comprising members of all major political parties and a wide range of Puerto Rican civil society, including representatives of the Vieques community. The Commission's June 1999 report demanded the end of all military activity on Vieques as well as the end of military control of the land. By July 1999, the pro-statehood governor backed down from his previous, pro-Vieques position; he "dissolved the Special Commission on Vieques and created a working group to negotiate—behind the backs of the previously participating sectors—with White House and Defense Department officials" (García-Muñiz 2001).

Details of the Rosselló-Clinton Agreement

In late winter 2000, Governor Rosselló reached an agreement with the Clinton administration that would allow resumption of naval training exercises on Vieques. Despite previous declarations that not one more bomb would fall on the island, the governor signed compromise legislation that allowed for limited bombing practice in exchange for a plebiscite. The plebiscite, scheduled for November 2001, would allow Viequenses to vote on one of two options: (1) that the Navy would cease all training no later than May 1, 2003; or (2) that training would continue and could include live fire exercises. No matter which option was chosen, the Clinton administration would request $40 million from Congress to support community development efforts on Vieques. If local residents voted to continue Navy training, the federal government would provide an additional $50 million.[13]

The Clinton-Rosselló agreement broke the rare consensus among Puerto Rico's political parties that bombing must end and had the distinct potential to derail the anti-Navy struggle. Significantly, the unity of the clergy in opposing continued bombing was essential in maintaining the mobilization at this point. In fact, Puerto Rico's religious leaders were responsible for what may have been the largest demonstration in the island's history when, on February 21, 2000, in San Juan, somewhere between 85,000 and 150,000 islanders protested a resumption of bombing. Nevertheless, the San Juan–Washington agreement signed by Governor Rosselló directed the Puerto Rican government to dismantle protestors' encampments on the Vieques target range. This would pave the way for a resumption of military maneuvers. In May 2000, the protestors were evicted from their year-long vigil on the bombing range. Yet the activists, who against all odds had created a nonpartisan movement, refused to accept the terms of the Clinton-Rosselló pact.

Indeed, Pedro Rosselló's perceived capitulation to Washington at least partly explains his statehood party's loss in the gubernatorial race of November 2000 to Sila Calderón, a *popular*, who pledged to get the Navy out of Vieques. Calderón not only rejected the Clinton-Rosselló agreement, but went further, saying that even three more years, until May 2003, was too long to wait for the Navy's exit. The statehooders also lost the mayoral race on Vieques and the new mayor, *popular* Damaso Serrano went to jail soon after his election for leading an act of civil disobedience on the bombing range.

One of Sila Calderón's first acts as governor was to remove the Commonwealth riot squads from Vieques on January 5, 2001, and leave only a group of local police to monitor the range. Also encouraging to the protestors was that on his last full day as president, Bill Clinton sent a directive to his Secretary of Defense requesting that the Navy find an alternative to Vieques.

In an introductory burst of goodwill, the incoming Bush administration agreed to postpone the March 2001 training exercises until various medical test results became available, although the military would not necessarily be restricted by the findings. At the time, island scientists were conducting health studies to see if there was a relationship between naval training and cancer, heart disease, and infant mortality among Viequenses.

Predictably, the Navy and the protestors would have different interpretations of findings on almost any health or environmental analysis. Governor Calderón characterized the new round of training exercises at the end of April 2001 as a betrayal of a January 2001 agreement she had signed with the U.S. Secretary of the Navy. That agreement called for the Navy to halt the bombing until the Department of Health and Human Services had reviewed a health study ordered by the Clinton administration in January 2001. By mid-April the Pentagon dismissed the health study on a variety of grounds and also dismissed the agreement with Governor Calderón, claiming that military readiness took precedence over other concerns.[14]

By April 2001, when the Navy planned to resume bombing, Vieques had turned into a national *cause célèbre*. New York Governor George Pataki traveled to the island to speak out against naval bombing exercises. Congressman Luis Gutiérrez, environmental lawyer Robert Kennedy Jr., Rev. Jesse Jackson's wife, Jacqueline Jackson, and New York political activist Rev. Al Sharpton were arrested and jailed for trespassing on Navy land there. Entertainers Marc Anthony, Ricky Martin, and José Feliciano, and athletes Tito Trinidad and Chichi Rodríguez signed an anti-bombing appeal to President Bush that appeared in the *New York Times* on April 26, 2001. Between April 1999 and April 2001 several hundred people were arrested for participating in protests, but in spring 2001 the numbers continued to grow with both prominent mainlanders and local Puerto Ricans spending time in jail (*San Juan Star* 2001.)

The great attention over Vieques is not hard to understand. New York politicians were jockeying for Latino votes in the 2001 mayoral race and the 2002 gubernatorial election. As even President George W. Bush, a former governor of Texas, understood, Vieques illuminated the growing national importance of Latinos as a political bloc. Vieques was not just a "Puerto Rican issue," but a struggle with salience for Latino voters throughout the United States. This understanding, no doubt, underlay Bush's dramatic June 14 announcement from Goteborg, Sweden that all bombing would stop by May 2003.

For the activists, though, two more years of naval training was too long, and the Vieques protests continued with the June 2001 resumption of maneuvers. Also, Governor Calderón organized a symbolic, non-binding referendum on July 29 of that year for Vieques residents. The local referendum produced a high turnout of eligible voters (80.6 percent), and the results were not surprising. Sixty-eight percent demanded an immediate exit by the Navy, while 30 percent voted to allow the Navy to stay indefinitely. The remainder (1.7 percent) opted for the Bush proposal allowing the military to remain until May 2003 (*New York Times* 2001).

On September 11, 2001, terrorist attacks in New York and Washington dramatically changed the political landscape. The themes of military preparedness and security became national priorities. Vieques activists sensibly called a moratorium on protests and the Puerto Rican government, like numerous other governments worldwide, expressed condolences and support for U.S. anti-terrorism measures (*New York Times* 2001.) The protest moratorium was an expedient strategy, since island activists found that their mainland support had evaporated, at least in the short term (*New York Times* 2001)

For more than a year after the September 11 attacks, it was not clear that President Bush would keep his pledge to have the Navy depart by May 2003. The issue became especially knotty in December 2002 when Congress voted to give the Navy and Marines the final decision on leaving Vieques and also cancelled a planned referendum for Viequenses to vote on the Navy's future on the island. Further, Congress ruled that the Navy had to certify that it had found a suitable alternative training site and that the federal government would retain the lands (15,000 acres comprising Camp García and other training grounds) rather than return it to the *municipio.* Still, President Bush continued to assure Governor Calderón that the U.S. military would leave as planned in May 2003; and in February 2002, Navy Secretary Gordon England wrote to the governor stating that he was personally opposed to continuing the Navy's presence beyond spring 2003.

Although the navy resumed training in April 2002, by August, twenty-two members of Congress were willing to separate the war on terrorism from Vieques and demanded that the president issue an explicit executive order. In September, Governor Calderón followed up with a letter to President Bush, and near the end of October Secretary England confirmed the Navy would leave; after sixty years, the U.S. Naval presence on Vieques ended on May 1, 2003. While the military's departure is viewed as a victory by most Viequenses, it is also seen as only the first success in the larger Vieques struggle. Activists refer to the three "Ds" in their present fight: decontamination, development, and devolution (return of the former military lands that were turned over to the Department of Interior, rather than to the *municipio*). It is to these interrelated issues that we now turn.

The Conversion and Cleanup of Vieques

On May 1, 2001, the Navy turned over 8,100 acres of land it had used on Vieques to various local and federal entities. President Clinton had issued an executive directive in January 2000 that instructed the Navy to return all 8,100 acres of the former Naval Ammunition Facility (NAF) on the western side of the island to the government of Puerto Rico. The debate over the nature of the western land transfer foreshadowed the larger debate over land use and cleanup that is occurring now, after the Navy's complete departure.

The agreement implemented on May 1, 2001 essentially achieved the objectives of the proposed "Vieques Land Transfer Act of 1994," originally submitted to Congress by then Resident Commissioner Carlos Romero Barceló. In the agreement concerning the 8,100 acres, 4,200 acres were given to the Vieques municipality, 3,100 acres were given to the U.S. Department of Interior, and 800 acres were given to the Puerto Rican Conservation Trust, a nonprofit group that maintains land in the public interest (Congressional Research Service 2004). The

Navy retained 263 acres for its radar facility as well as a 12.1-acre right-of-way. Before the May 1 transfer, some environmentalists urged the Vieques Mayor, Damaso Serrano, not to sign the agreement for his 4,200 acres because it had no provision for the massive remediation effort that would be necessary (*San Juan Star* 2001).

Mayor Serrano signed the land transfer agreement, and concerns over cleanup remain at the heart of efforts for present and future planning for Vieques. Detailing the different military uses of the western and eastern parts of Vieques will permit understanding the depth of the cleanup challenge that is the Navy's legacy.

Land on the western side of the island was used primarily for ammunition storage and was also the site of a small operational base. The western part of Vieques could be particularly valuable in the future because it represents the closest transportation point between Vieques and the main island. At present, though, the area needs serious environmental attention. Nearly two million pounds of military and industrial waste—oil, solvents, lead paint, and acid—and other refuse were dumped in mangroves and sensitive wetlands areas. Unexploded ordnance may also exist there.

In the transfer agreement, the Navy was charged with cleanup according to land use, but this is not a straightforward process. Indeed, part of the evolving struggle in western Vieques now centers on land use designations. Several land use categories exist, and land designated for residential use, for example, would have to be cleaned up much more thoroughly than land to be used for conservation purposes—land that would not be lived on.

Of the 8,100 western acres transferred in 2001, the U.S. EPA and the Puerto Rico Environmental Quality Board deemed seventeen sites in need of further study because of possible contamination. By May 2003, scientists had decided that nine of the seventeen sites were not seriously contaminated. The eight remaining sites remain under investigation (Atlantic Division, NAVFAC 2004). There is special concern about the former Open Burn/Open Detonation area, in the western part of the island. This site was used for disposing of leftover and defective ordnance; old munitions, bomb components, and flares were burned there in an open pit. The site was closed in 1970 after an accident involving three youths, but unexploded ordnance may still exist in this area (Márquez and Fernández 2000; UMET et al. 2000). Tests from the western side Resolución aquifer, however, show heavy metal contamination coming from the ordnance sites on the Interior Department land (*San Juan Star* 2001).

It is no surprise, then, that the 3,100 acres given to the Fish and Wildlife Service have become the first part of the Vieques National Wildlife Refuge, thus barring Viequenses from much of the returned land and also sparing the Navy the expense and effort of a thorough cleanup. It is now a common practice throughout the United States for the Pentagon to transfer polluted former base land to the U.S. Fish and Wildlife Service for use, paradoxically, as "wildlife preserves."[15]

Although contaminated, the land in western Vieques has not suffered the severe ecological destruction from six decades of bombing that is the case in the east. A thorough cleanup of the eastern part of the island will be much more dramatic in scope than in the west. The eastern side of Vieques was used for bombing exercises and maneuvers from the 1940s to the time the Navy left in 2003. The cleanup of firing ranges has proven one of the most dangerous, expensive, and challenging tasks in the military base conversion process (Sorenson 1998). That is why the entire eastern side of Vieques, consisting of 14,699 acres, was transferred to the U.S. Fish and Wildlife Service when the Navy departed in May 2003.

According to the Navy, eastern Vieques has been bombed an average of 180 days per year. In 1998, the last year before protest interrupted maneuvers, the Navy dropped 23,000 bombs on the island, the majority of which contained live explosives (U.S. Navy 1999). The most intense destruction was in the Live Impact Area, which constituted 980 acres on the island's eastern tip. This has now become a National Wilderness Area with human presence prohibited. All 14,699 acres and the surrounding waters of eastern Vieques had been used as shooting ranges, amphibious landing sites, and toxic waste dumps since the 1940s. Coral reefs and aquatic plants sustained significant damage from bombing, sedimentation, and chemical contamination. Nitrates and explosives have contaminated the groundwater (Márquez and Fernández 2000; Rogers, Cintrón, and Goenaga 1978).

The cleanup of unexploded ordnance on land is a clear safety issue and would be a top priority. Of particular concern are revelations that the Navy has fired depleted uranium munitions on the range, because of the risks this may pose for the civilian population.[16] Numerous unexploded bombs remain off the shores of Vieques; cleaning offshore must be part of the long-term cleanup effort.

At least three economic conversion and development plans already exist to promote Vieques's future, including plans for an environmentally sensitive, sustainable tourism industry (GAPT 2000; McIntyre and Dupuy 1996; Rivera Torres and Torres 1996). The obvious problem is that all future plans presume an island free of environmental hazards, a presumption that requires a major financial commitment and act of political will from the federal government. Indeed, the chances that the Navy and EPA under current statutes would be involved in a full cleanup of both the western and eastern parts of the island are slim at best.[17]

A major step forward in the Vieques struggle came in February 2005 when the EPA formally designated Vieques a Superfund site. The process had begun almost two years earlier when then Governor Sila Calderón had requested the island's inclusion on the National Priorities List (NPL) of most hazardous waste sites. Specifically, the NPL designation requires the Navy to remediate the Atlantic Fleet Weapons Training Area on eastern Vieques as well as waters and cays in and around the island. Although Governor Calderón also had requested inclusion of Culebra as part of the Superfund site, it is likely to be cleaned up under another program (the Formerly Used Defense Sites) run by the Army Corps of Engineers.

Still, the cleanup will not be short. To understand the probable duration and extent of this environmental justice struggle, it is helpful to consider two relevant cases. After the Navy left Culebra in 1975, more than two decades elapsed before funds were allocated to clean up ordnance. This limbo period witnessed widespread land speculation, gentrification, and economic marginalization of the local community (Iranzo Berrocal 1994; Rivera Torres and Torres 1996). A second case is Kaho'olawe, Hawaii, where the Navy also had a live impact range. In 1990, President George H. W. Bush issued an executive order to end the bombing exercises on that uninhabited island, which lies seven miles off Maui. In 1993, Congress agreed to finance a ten-year, $400 million cleanup effort for the forty-five-square-mile island. It took five years for the process even to begin, and rancor existed throughout between the Navy and Hawaiians. By 2000, the Navy had cleaned up only one-tenth of the island (Klein 2001). In April 2004, Kaho'olawe was turned over to the State of Hawaii with 77 percent of surface munitions and 9 percent of subsurface munitions cleaned; almost all of that island remains off–limits to civilian use. The failure of cleanup efforts in the state of Hawaii is particularly troubling because Puerto Rico, a U.S. territory, lacks the political leverage of one of the fifty states.

Finally, to put the Vieques struggle in a larger context, the long-term environmental cleanup of military bases is still highly contentious within the United States. For decades, the Department of Defense (DOD) has relied on national security concerns to argue for exemptions from environmental legislation. In the 1980s alone, the U.S. military was estimated to have generated 500,000 tons of toxic waste per year, more than the top five U.S. chemical companies combined. One report identified 20,000 sites at 1,800 military installations with varying levels of contamination. Nearly 100 of these sites would warrant placement on the National Priorities List of the Superfund cleanup effort (Renner 1994). It is notable that in the post–9/11 priority on security, the Pentagon has successfully argued for exemptions from parts of some environmental legislation.

Conclusion: Environmental Struggles and the Deepening of Democracy in Puerto Rico

The end of the Cold War opened a political space in which Vieques's long-simmering grievances against the Navy could be expressed. For decades, opposition to the military in Puerto Rico was perceived as anti-colonial and anti-American. Legitimate grievances about Vieques' stifled economy and environmental damage from bombing became mired in cold war politics. The collapse of the Soviet Union created a new context.

A focus on health, environmental protection, and human rights are key elements of the revitalized movement's efforts to expand and reach new constituencies. The Vieques struggle fits nicely into the environmental justice framework that gained a degree of national legitimation and institutionalization in the 1990s.[18] The basic theme of the environmental justice movement is that the poor and ethnic minorities suffer disproportionately the burden of environmental risks taken by industrial society. Environmental concerns thus expand beyond technical discussions to become issues of civil and human rights.

After the events of September 11, 2001, Vieques activists observed a moratorium on anti-Navy protests for a brief period given the nation's sense of mourning and the move to a war footing. Within a month, however, activists resumed their struggle, and by August 2002 major politicians were again able to embrace the issue.[19]

That the military exit became inevitable despite the U.S. war footing is testimony to the resilience of the Vieques movement. Now that the grassroots organization is in place and the struggle has been framed in terms of human rights and environmental justice, the health and environmental concerns that surfaced in the late 1990s will continue as issues even with the Navy's departure. The Vieques struggle has contributed to a deepening of democracy in Puerto Rico. We hope it will serve as a model for other grassroots environmental justice struggles in Puerto Rico, and will thus allow citizens to move beyond the paralyzing divisiveness of traditional party politics to participate on issues of significance in their daily lives. Such grassroots groups have strengthened civil society, a key to a smoothly functioning democracy. The Vieques struggle represents a quest to end a legitimate grievance without having to choose a status option; the large majority of Puerto Ricans clearly rejected what analyst Juan Manuel García Passalacqua called "cupones por megatones," or the federal government's policy of providing welfare benefits in exchange for holding naval maneuvers.[20]

The environmental struggle in Vieques may also have contributed to incisive questions about the quality of democracy in the United States. Indeed, the issue of military responsibility for environmental contamination increasingly cuts to the question of civilian control over the military, a basic tenet of a democratic society. After decades of secrecy surrounding its activities, the military is emerging as the single largest polluter in the United States, having produced 27,000 toxic waste sites in this country (*Environmental News Service* 2001; Sorenson 1998, 78). The military, protected by the rhetoric of national security, has not been held fully accountable for its toxic legacy. Therefore, while an end of bombing on Vieques represents a clear victory for Viequenses, activists will need organizational skill and perseverance to continue the struggle to clear the Navy's legacy of contamination.[21] The next stage of the struggle for decontamination and development will not be brief, but it will enhance the quality of life and democracy in Vieques and Puerto Rico.

Notes

1. According to Carmen Ortíz Roque of the Puerto Rico Surgeons and Doctors Association, the infant mortality rate in Vieques has climbed in the past twenty years while

decreasing in Puerto Rico as a whole. Between 1990 and 1995 infant mortality rates were 50 percent higher in Vieques than in Puerto Rico as a whole (*El Nuevo Día*, 23 February 2000). Puerto Rican Governor Sila Calderón publicized a study that suggested that residents suffer from vibroacoustic disease, an unusual heart disorder associated with exposure to loud noises like jet engines or deep explosions (*New York Times*, 14 January 2001). The study was later challenged by Johns Hopkins researchers (*New York Times*, 15 July 2001).

2. See for example *National Catholic Reporter*, 21 March 2000.

3. The "framing" notion comes from Goffman 1986.

4. Keck's focused analysis on Acre is found in Keck 1995. A more general discussion of this work is in Keck and Sikkink 1998.

5. Only two other Puerto Rican municipalities—San Isabel, dominated by the Aguirre Sugar Company, and Guanica, dominated by the South Porto Rico Sugar Company—had sharper inequalities of land ownership (Ayala 2001).

6. For further discussion of the effect of the military expropriation of land on Vieques residents see McCaffrey 2002, chaps. 1 and 2; and Ayala 2001.

7. For further discussion on the plight of the Bikini islanders see Delgado 1996; Dibblin 1988; Kiste 1974; Weisgall 1994.

8. For further discussion of the Culebra movement see Delgado Cintrón 1989 and McCaffrey 2002, chap. 3. For discussion of the growing militancy of the Puerto Rican independence movement and the struggle against the draft see Nieves Falcon, García Rodríguez, and Ojeda Reyes 1971.

9. Angel Rodríguez Cristóbal was found dead in his prison cell on November 11, 1979, two months into a six-month term. Prison officials declared the death a suicide, but an independent autopsy performed by the family concluded that he was beaten to death. Photos of the cadaver showed that the face was heavily bruised, inconsistent with a finding of suicide by strangulation.

10. See McCaffrey 2002, chap. 6.

11. The ability of environmental issues to rally Puerto Ricans regardless of partisan ties is documented in Baver 1993. Also, sociologist Myra Muñoz, for example, has examined at least one hundred struggles since the 1970s in which Puerto Ricans have crossed party lines and banded together on issues. See Muñoz 2001.

12. The large presence of women is common in environmental justice struggles in the United States. On this point see Harvey 1999, 153–85.

13. "Solution on Vieques Takes a Step Forward," *New York Times*, 29 February 2000.

14. See "Navy Bombing is Betrayal, Puerto Rican Governor Says," *New York Times*, 29 April 2001; the Navy's position is found in Jack Spencer, "The Importance of Vieques for Military Readiness," *The Heritage Foundation Backgrounder*, Washington, D.C. No. 1411, February 16, 2001.

15. Sorenson (1998, 82 n. 168) notes that most of the 50,000 acres of the most contaminated firing ranges have been transferred to the Department of Fish and Wildlife.

16. For a discussion of the depleted uranium controversy see links to depleted uranium on www.viequeselibre.org and visit the website of the Military Toxics Project at www.miltoxproj.org.

17. "2 Experts Testify on Cleanup of Vieques," *San Juan Star*, 22 July 2001. In this article, the experts were a representative of the Center for Public Environmental Oversight and the former chief of the Army's Environmental Law Division. Testimony was given before the Puerto Rican Senate's Agriculture, Natural Resources, and

Energy Committee, which was holding hearings on environmental issues in Vieques. See Also Shulman 1992. Quoted in Switzer 2001, 131 n. 27.

18. In 1994, for example, President Clinton issued an executive order stating that federal agencies must consider principles of environmental justice in their decision making. A useful discussion is in Harvey 1999, 153–85; or in Agyeman, Bullard, and Evans 2003.

19. E.g., "Vieques Issue Is Put on Hold in Response to Terrorism," *New York Times*, 27 September 2001, reveals the early unwillingness of both protesters and politicians to push their demands, but by early 2002, at least island activists and politicians began to speak out again.

20. García Passalacqua has used this phrase numerous times in recent years; one example is "Calderón Is Walking the High Wire on Vieques Issue," *San Juan Star,* 25 March 2001.

21. Viequenses searching for models of community-based sustainable development might turn to the Salinas case as detailed in Berman-Santana 1996.

PART IV

Risky Environments and the Caribbean Diaspora

CHAPTER 9

Environmental Justice for Puerto Ricans in the Northeast

A PARTICIPANT-OBSERVER'S ASSESSMENT

RICARDO SOTO-LOPEZ

Introduction

In recent decades, Puerto Rican community activists in New York and the northeastern United States and environmentalists from Puerto Rico have entered into a dialogue on the quality of environmental protection afforded our geographically dispersed community. Community activism around environmental protection has a forty-year history in Puerto Rico (see, e.g., García-Martinez and Valdés-Pizzini, this volume). Activism in Puerto Rican communities in the northeast stems from the perception that environmentally undesirable facilities are disproportionately located in communities that have been predominantly Puerto Rican and now also have newer Latin American immigrants; at the same time, these communities have witnessed a decline in the health of their residents.

In the past fifty years, Puerto Ricans on the island and in the Diaspora have experienced environmental degradation comparable to the process occurring 150 years ago in the continental United States. Since the 1950s, Puerto Ricans have felt the impacts of rapid industrialization on the island and de-industrialization in the northeast, two distinct processes that have left in their wake contaminated communities suffering from a range of occupational and community health problems. These problems, which have further impoverished the poorest segments of the Puerto Rican population, are due largely to siting of polluting facilities disproportionately in the regions and neighborhoods where Puerto Ricans live

The roots of the environmental problems now faced by Puerto Ricans can be traced to the movement of capital and the economic restructuring of the industrial northeast and in Puerto Rico after World War II. The economic restructuring in the island—widely known as Operation Bootstrap—was implemented through the Industrial Incentives Acts of 1947 and 1948, pieces of legislation initiated by the Puerto Rican government ostensibly to promote economic development and jobs. While Operation Bootstrap resulted in a dramatic increase in per capita income and catalyzed structural change in Puerto Rico's largely agrarian society, it did not generate enough employment to absorb all of the labor pushed out of the declining agricultural sector. For this reason, Operation Bootstrap incentives impelled a mass migration of Puerto Rican agricultural workers to the United States. These migrants established communities throughout the northeast, but primarily in New York City, where many found jobs in the city's declining manufacturing sector.

The survival of low-wage manufacturing in New York was a primary reason for the mass migration of Puerto Ricans in the period following World War II. Unlike the economies of large midwestern cities, which were based on a few large-scale industries such as steel and auto manufacturing, the economy of New York City featured a manufacturing sector that focused largely on the production of garments and other non-durable goods.[1] Although the city was a major manufacturing center, as in other cities of the northeast, manufacturing was scattered among many small firms.

Firms producing non-durable goods in the cities of the northeast were subject to little regulation in the early postwar era (1947–1952), when the Puerto Rican migration was at its peak. Puerto Ricans entering into the labor market in the postwar years were concentrated in low-wage jobs that often entailed exposure to hazardous substances in the workplace. As a result, they faced greater health risks than did the U.S. population as a whole. Puerto Ricans continue to face environmental health risks in the workplace. The use and disposal of hazardous substances continues to affect the health of large numbers of Latinos who work in occupations like textiles, auto repair, health care, metal work, and printing where exposure to hazardous materials is the norm (Pellow 2002).

Land Use Policy and Exposure to Pollution

Hazards in the workplace are compounded by pollution in Latino neighborhoods. The proximity of residential areas to industrial sites compounds the health problems faced by Puerto Rican communities. This problem can be traced to the unplanned nature of late-nineteenth-century industrial and residential development in New York City's boroughs and to the neglect of industrial siting issues in subsequent zoning efforts. Modern zoning laws governing land use in New York City date to 1916. Zoning ordinances introduced in the early twentieth century were designed to protect prime real estate in Manhattan. These municipal laws initially regulated the height of structures and land uses within designated zones; they neither banned nor phased out preexisting, incompatible industrial uses in the mixed-use neighborhoods that would be settled by Puerto Ricans. Small and mid-sized firms of the type found in Puerto Rican neighborhoods are among the largest users of hazardous substances in their manufacturing processes, and are significant producers of hazardous waste. Thus, in the 1940s and 1950s, the Puerto Rican community's dependence on the most contaminating industries, and their settlement patterns in relation to these industries, meant high levels of exposure to environmental health risks.

In the 1960s, Performance Standard Zoning set limits to adverse off-site impacts of odor, noise, and signs. This was prior to the passage of modern, scientifically based environmental protection statues and regulations, which set goals for environmental quality as well as for the protection of human health.[2] These goals have been achieved either by banning certain substances or by setting ceilings on permissible emissions and concentrations of designated pollutants.

At present, New York City's Department of City Planning promulgates zoning and land use ordinances and conducts hearings for the approval of proposed projects and zoning map changes. Actual compliance with zoning regulations is enforced by the Buildings Department. While the Department of Environmental Protection's regulations are referenced in certain performance standards in the City's Zoning Resolution, evidence from Hunts Point and Greenpoint-Williamsburg suggests that there is little if any coordinated enforcement of these standards. The same is true of most major northeastern cities where Puerto Rican communities are concentrated. The result of this legislative and regulatory dysfunction is a concentration of health risks and, more generally, an unpleasant environment in many low-wage Puerto Rican and Latino communities throughout the northeast.

Manufacturing has been an important source, but not the only source, of pollution in cities where Puerto Ricans live. In 1984, an estimated 13,000 hazardous waste generators were operating inside New York City limits (New York State Department of Environmental Conservation 1984). Gasoline service stations, motor vehicle repair shops, dry cleaners, electroplating, photo labs, printing and dyeing operations are among the most common generators. Wastes from these facilities include lead and acid from used batteries, cleaning and degreasing solvents, heavy metal sludge, paints, and inks. While in the late 1980s and 1990s serious efforts were made to encourage a shift to the use of more environmentally friendly products and processes, the health of Latino communities in New York and the northeast is still compromised by contamination from dirty power plants and lead residues (Sierra Club 2004).

In New York City, until recently, hazardous waste disposal occurred without significant government scrutiny. Some industries sent their waste materials to city-owned landfills. As García et al. note, this is also the case in Puerto Rico (chapter 6). Other firms have stockpiled wastes indefinitely on their premises, creating toxic waste sites. Not infrequently, industrial wastes have simply been abandoned on vacant lots. This has occurred in the Puerto Rican communities in the South Bronx, the Williamsburg section of Brooklyn, in Bridgeport, Connecticut, and in Newark and Elizabeth, New Jersey.

The presence of hazardous materials on a site may not be obvious. Sites that appear to be clean and have no commonly known sources of contamination may have been affected by the use of toxic materials on the site or in surrounding areas. Also, the soils and groundwater beneath industrial sites in the northeast have often been contaminated. When firms use, store, and handle hazardous materials, contaminants often migrate away from the site and into the groundwater. At other times, hazardous materials may be incorporated into buildings and structures on the site, as in the case of lead paints or asbestos insulating, tiling, and roofing materials. The toxic burden is particularly heavy in older industrial neighborhoods.

When I returned to my old neighborhood in the South Bronx in the mid-1980s as a New York City Planner working with local redevelopment efforts, I found what I had expected—abandonment of the residential sections of Mott Haven, Longwood, Hunts Point, Morrisania, Claremont, Crotona Park East, Bathgate, and Melrose. However, I was not prepared for the extent of abandonment of the industrial areas. Boarded up industrial buildings were more difficult to understand than those vacant, burned-out residential buildings.

What I saw in these neighborhoods were the effects of depopulation and de-industrialization, which had resulted in the loss of hundreds of thousands of industrial-sector jobs since 1965 and the physical decline and destruction of buildings and infrastructure. In the place of industries I found an agglomeration of polluting waste-transfer facilities that discouraged productive land use. I also found concentrations of so-called recycling facilities in close proximity to one another. Neither the technologies nor the practices used by these firms were environmentally sound (*New York Times*, 1994). Paradoxically, it was federal environmental protection policies that had contributed to the clustering there of waste-transfer stations and to the clear example of environmental injustice. Specifically, the use of precious industrial land for siting municipal waste processing facilities enabled the city to comply with a federal mandate to stop ocean dumping. Finally, I encountered plans for reactivation of generally less polluting infrastructure, as in the case of the Harlem River Rail Yards. While seemingly positive moves, these still may have negative impacts on neighborhoods with significant Puerto Rican and other low-income populations, not only in New York City but in other northeastern cities as well.

Participation in Land Use Decision Making

As the Puerto Rican/Latino community struggles to rebuild its neighborhoods and to fend off projects that hinder its physical, social, and economic progress, its active participation in land-use decision making becomes essential. It is unfortunate, then, that the forms of stakeholder participation that prevail in local land-use planning discussions are more formal than substantive. Latino neighborhoods are encouraged to engage in passive participation processes that are politically controlled: the "impacted," "host," or "targeted" community has to respond to projects that others bring in. When local people have raised concerns about development efforts, the response by municipal authorities has generally been to appoint them to Citizen's Advisory Committees organized by government agencies, elected officials, and the private sector. Sadly, the experience has been that once activists have been placed on these committees, community mobilization to protect itself against a project's deleterious environmental impacts all but vanishes. Alternative structures—coalitions, ad-hoc committees, neighborhood-based organizations—are needed to develop proactive strategies that bring citizens to the negotiating table at the beginning of the process. These people must be seen as legitimate stakeholders who have the right to help make decisions regarding the use of neighborhood land because these decisions will impact the quality of their lives and the community's economic development.

Making decisions about land use is essentially a political process. The process can be demystified by ensuring local participation in it and by identifying points in the process where the community can make a political intervention. This in turn requires access to information and knowledge about community land use and the development of community-based plans.

Success Stories

The history of the environmental justice movement in the United States usually starts with either the neighborhood movement at Love Canal led by Lois Gibbs in 1979 or with the NAACP-led protests against the dumping of PCBs in Warren County, North Carolina in 1982. Yet a case can be made that the Young Lords in the late 1960s were the first in the United States to link the problems of poverty, racism, and pollution (Gandy 2002). The Lords' early demands for better sanitation and health care in East Harlem and the South Bronx served as a model for the later struggles of New York's Puerto Ricans against pollution in the 1990s. Groups like Brooklyn's El Puente and the South Bronx Clean Air Coalition owe their heightened consciousness of their right to control their community space to the legacy of radical environmentalism left by the Lords in New York City.

In a number of cases, community participation has worked to the benefit of the Puerto Rican neighborhoods faced with potential environmental hazards. One success story took place in Sunset Park, Brooklyn. In 1993, in response to pressure from the Latino and Asian communities, city politicians moved to block a plan to site a sludge composting plant in the neighborhood. Brooklyn Congresswoman Nydia Velasquez collaborated with the Bronx Borough President at that time, Fernando Ferrer, to "can the plan." This effort was unprecedented.

Another example involved the organization *Nos Quedamos* (We're Staying), in the Melrose neighborhood of the Bronx. In this case, the fight took ten years, but residents were ultimately successful. In 1984, a Bronx development plan initiated by the City Planning Department, championed by the Bronx Borough President, and presented to the City Planning Commission would have displaced longtime Melrose residents, predominantly Latino families. *Nos*

Quedamos succeeded in changing the development plan to protect their homes and businesses and ended up playing a major role in the planning process. They were able not only to avoid displacement but also to reshape the plan to accommodate their vision for land use and a more sustainable environmental orientation than had been the case in the initial project.

A third major case involved the Brooklyn Navy Yard, where the organizational clout of the Latino community on environmental planning at first seemed doubtful. This struggle underscored the need to revamp the routine processes of environmental impact assessment, site analysis, and state and city environmental quality review, and instead turned to a cumulative environmental impact analysis. With the withdrawal of the federal government from the Navy Yards in 1966, much of the site reverted to New York City. A plan for a giant waste incinerator at the Yard was first proposed in 1979. In 1994, elevated levels of toxic chemicals were discovered in the soil and groundwater in a section of the 92-acre area; the presence of these toxic chemicals was said to be the result of 150 years of shipbuilding on the site. However, no New York State or City official could explain how and why the contamination had gone undetected during the fifteen-year period when the plan for the incinerator was under consideration. An Environmental Impact Assessment, conducted about ten years prior to the discovery, made no mention of such toxic waste on the site.

However, Latinos and Hasidim in the Greenpoint and Williamsburg neighborhoods in Brooklyn joined in a highly unusual alliance to defeat the planned transformation of the derelict Brooklyn Navy Yard. This alliance—which owed much of its strength to the efforts of El Puente, a Latino community organization—was so unusual that when Puerto Rican community leader Luis Garden-Acosta first went to meet with Hasidic Rabbi David Niederman of United Jewish Organizations, he felt "like Nixon going to China" (Gandy 2002; Checker 2001). The incinerator plan was ultimately defeated in 1995, when the Latino and Hasidic communities, with support from other activists from the Fort Greene and Red Hook neighborhoods, persuaded the City Council to abandon it.

Around the same time, contestation over waste disposal occurred when the Bronx Lebanon Hospital Medical Waste Incinerator was built in the Bronx on a private site with public monies. In this case, a voluntary consortium of hospitals and a waste management corporation used New York City Industrial Development Agency bonds to construct the incinerator. Facilities that are developed with public dollars are required by law to enhance employment opportunities in the low-income neighborhoods in which they are built. The incinerator was deemed to be "environmentally beneficial" and an "employment generator." Unfortunately, the community was not initially privy to the permitting review process for the incinerator, even though that process was the community's only opportunity to formally protest the siting of the plant.

To make matters worse, in 1994, the New York State Department of Environmental Conservation promulgated rules that would minimize public participation in reviews of state permits for the operation of this type of facility. Thanks to its extraordinary reach and painstaking labor, the South Bronx Clean Air Coalition uncovered detailed information concerning the environmentally degrading practices used by the proposed waste management corporation in its operation of similar facilities in other low-income communities and communities of color. Ultimately, the medical waste incinerator was closed down. Yet the coalition's exhaustive efforts may be precisely why this public review process came under attack.[3]

A Good Project Defeated

The successful efforts described above were primarily reactive in nature. Neighborhood organizations responded to threats and prevented environmentally risky projects from moving forward. However, minority and poor communities stand to benefit more from proactive projects that combine neighborhood improvement with employment generation in sustainable enterprises. One such effort was begun in 1992, but was scrapped in 2000 when New York City Mayor Rudolph Giuliani withdrew city support. Despite its eventual failure, positive lessons can be learned from the effort, which entailed broad-based community participation in the planning process. This project brought together the Banana Kelly Community Improvement Association, the Natural Resources Defense Council, and a Swedish paper company in an initiative to develop the Bronx Community Paper Company in the South Bronx. New York City generates hundreds of thousands of tons of office waste paper a day; the goal of the project was to transform the city's "urban forest" of recyclable paper products into high-grade market pulp for sale to international paper companies. It was also hoped that the project would spur local economic development and generate new jobs.

The Bronx Community Paper Company project was considered a model for sustainable development. It was predicated on principles of environmental justice in that it sought to revitalize a "brownfield" site located in the Harlem River Rail Yards. It attempted to show that job production, socioeconomic development, private-sector profitability, and environmental protection could work in a mutually supportive fashion. It is an example of how a facility should be sited and designed so as to minimize negative environmental impacts. It is the type of development we all should be searching out for our communities and our island(s). Unfortunately, due to a variety of factors—the project's lack of economic feasibility for the corporation, the utopian ideals of the planners, the political hostility of Mayor Giuliani, and infighting among Bronx community groups—the project fell apart in 2000 (Hershkowitz 2003; Harris 2003). The project's failure points to a serious obstacle to environmentally sustainable development in Puerto Rican neighborhoods, and in poor urban neighborhoods more generally. Local infighting is a not uncommon feature of environmental justice struggles, and it perpetuates the disempowerment of low-income and minority residents (Roberts and Toffolon-Weiss 2001).

A Framework for Community Land Use
and Environmental Justice Analysis

These and other examples of environmental justice efforts in Puerto Rican communities suggest that an analytical framework for the land use aspects of environmental justice struggles is useful and is a crucial first step toward a comprehensive understanding of the threats confronting poor communities. Such a framework can help communities to identify critical social, economic, and political factors that may aid in developing the public policy approaches tailored to their specific needs. It would provide a geographical and spatial analysis to define the geographic boundaries of the community, identify settlement patterns, and describe the spatial relationships between the residential, industrial, and commercial areas in the community. It would ask about concentrations of industries and/or public and private solid-waste facilities nearby. It would also identify dilapidated housing with high lead and asbestos levels. This spatial analysis would look at the ways in which settlement patterns and land use patterns are superimposed upon one another at present, and how these would change in the future given development trends. It would then ask about the environmental benefits, risks, and hazards that are likely given these patterns and trends.

A second component of an analytical framework would be socioeconomic. What is the current socioeconomic condition of the neighborhood? What is the nature of the local economic base? To what extent does the local economy exacerbate environmental problems in the community?

A third component would be to assess the history of and potential for community-based involvement in land use decision making. This entails not only an analysis of the capacity for community mobilization, but the presence of government bodies in the neighborhood. The first question to be answered is who makes land use decisions for the community? What use is made of the formal public participation mechanisms? What community groups have played a role in land use decision making and what have been the results? Finally, how have current public policies helped or hindered environmental protection in the community?

Fourth, the framework would include an assessment of the environmental impacts of installed and projected facilities on community health, economic development, and civil rights. It is important to get an idea of the health, economic development, and legal issues confronting the community as a result of environmental concerns. What recourse and resources does the community have to address these issues? Potential allies in a campaign to improve environmental quality in communities of color might include members of the business community, the clergy, health care workers, university students and faculty, and local unions. With this information in place, it becomes possible to draw implications about the effect of public policy advocacy and to develop short- and long-term community based plans, recommendations, and policies for land use.

Conclusion

There is a general recognition among Puerto Rican community activists on the island and in the northeast that environmental justice struggles must be led by those who are most affected by the environmental and health hazards associated with public and private development in their neighborhoods or from exposure to hazardous substances in the workplace (Grosfoguel 2003). There have been some successes to date, but the war is far from over. One waste transfer station may be defeated, but then a power plant moves in—as happened in New York City in 2001 (*New York Times* 2001).

Poor communities need to work with legal, scientific, and other advocates to address environmental justice concerns and to prepare alternative sustainable development policies. However, the initiative lies with the community. And "the community" may have to be defined more broadly so that promoting environmental justice in the northeast does not end up promoting injustice in poor communities elsewhere.

Notes

1. Non-durable goods manufacturing includes production of foods and kindred products; textiles; apparel; paper and allied products; printing and publishing; chemicals and petroleum products; rubber and plastic products; leather and leather products.
2. For example, the National Environmental Policy Act (NEPA) was enacted in 1970, and the Resource Conservation and Recovery Act (RCRA) was enacted in 1976. The rules and regulations resulting from these acts where promulgated into the late 1970s. New York's State Environmental Quality Act vas not passed until 1975. While New York City had issued an Executive Order in 1973 as a result of NEPA, the city's Environmental Quality Review was established only in 1977. The entire NYC Environmental Quality Review process, which governs environmental review of discretionary land use actions, was revised with the 1989 Charter. Importantly, Puerto Ricans generally did not benefit from these laws, since the rules and regulations were promulgated at least ten years after many of these industrial workers were displaced due to industry closings and relocations.
3. In 2000, the New York League of Conservation Voters reported that the Hunts Point, Concourse, Port Morris/Mott Haven, and Soundview sections of the Bronx, whose population was 66 percent Hispanic and 30 percent black, had facilities that were processing 40 percent of New York City's garbage. The area also had several medical waste incinerators and power turbines (www.nylcv.org/ecofiles/bronx). See also, "Bronx Loudly Opposes Waste Station Plan" *New York Times*, 9 March 2000.

CHAPTER 10

Environmental Risk and Childhood Disease in an Urban Working-Class Caribbean Neighborhood

LORRAINE C. MINNITE AND IMMANUEL NESS

Introduction

The environmental justice movement of the last two decades has confronted dimensions of poverty and racism previously overlooked in movements for social justice: the socially and geographically inequitable distribution of the costs of environmental degradation and pollution accompanying industrialization (Freudenberg 1984; Bryant 1995; Novotny 2000; Rhodes 2003). The historically uneven pattern of this distribution reflects a class and racial bias tied to the position of poor and working-class whites and racial and ethnic minorities in the capitalist economy. Their residential segregation creates opportunities for spatially disaggregating the costs and benefits of industrial production and other polluting functions of the local economy, resulting in the disproportionate concentration of environmentally hazardous activities in low-income and minority neighborhoods across the United States (United Church of Christ Commission for Racial Justice 1987; U.S. Environmental Protection Agency 1992; Goldman and Fitton 1994).

The decisions of businesses and governments to site noxious economic activities such as those producing high levels of air and water pollution, sewage treatment plants, toxic waste landfills, incinerators, and bus depots in or near low-income neighborhoods, usually where land values are cheapest, compound the multiple burdens of poverty (Been 1993; Chase 1993; Bullard 1994; Centner, Kriesel, and Keeler 1996). Moreover, environmental racism, or the deliberate targeting of communities of color for toxic waste facilities, contributes to higher incidence of poor health and disease among people least likely to afford quality health care (Collin and Collin 1997; Institute of Medicine 1999; Cole and Foster 2001).

In urban areas this form of discrimination is facilitated by a complex blend of political and economic forces that capitalize on existing segregated housing patterns, ghettoization, and minority disempowerment. In New York City, for example, toxic environmental hazards are more abundant in the neighborhoods of Mott Haven in the Bronx, Washington Heights in Upper Manhattan, and East New York and Sunset Park in Brooklyn, where the quality of housing is poor and large numbers of African Americans and people of Hispanic origin live. New York City's pattern of ethnically and racially divided neighborhoods diminishes the crisis of environmental decay for the general population of the city, since politicians and public health officials can isolate and ignore them without serious repercussions. This happens because many people living in these neighborhoods are usually inactive in or excluded from the political process. However, the environmental crisis in New York is not limited to politically excluded neighborhoods of poor housing quality. It is also important to understand the relationship between race, class, and de-industrialization. As the Puerto Rico–Northeast Environmental Justice Network stresses:

> Puerto Ricans have experienced the consequence of rapid industrialization on the island and de-industrialization in the Northeast that has left a legacy of environmental pollution and a range of occupational and community health impacts having the common effect of further impoverishing the community. (Puerto Rico–Northeast Environmental Justice Network 1995)

This chapter addresses some of the risks to health from environmental pollution in an urban, working-class, mostly Caribbean immigrant neighborhood. It analyzes data from a survey of low-income parents and their awareness of the risks from the environmental hazards they face. It also assesses the role of social capital in helping these parents protect their children's health. The next section briefly discusses how patterns of immigration and settlement can complicate the work of urban public health providers in addressing health information needs in low-income immigrant neighborhoods. Next we present findings from our survey of Caribbean Hispanic immigrant parents in a Brooklyn, New York neighborhood and explore how connections to community institutions like schools and churches are associated with levels of awareness of environmental hazards and the risks they pose to health. We conclude with support for an emerging family-community paradigm in public health that emphasizes building on the strengths of a community's assets in combating the environmental health risks facing the urban minority and immigrant poor.

Immigration, Cultural Diversity, and Health Care

De-industrialization and the transformation to a post-industrial economy in the United States has been accompanied by an expansion in immigration and cultural diversity. Population diversification through immigration is a complex phenomenon, but in its most far-reaching trends it reveals some singular features. One that has attracted much attention is the "Latinization" of the United States, with the number of people identifying themselves to the U.S. Bureau of the Census as Latinos or Hispanics now surpassing African Americans as the largest race or ethnic group among the nation's population (Ramírez and de la Cruz 2002). Another is the contribution of immigration to persistent levels of urban poverty, which is not to attribute the maintenance of urban poverty to the immigrants themselves, but rather to note the effects of observed patterns of settlement of the poorest and least educated immigrants, and the segregation of a disproportionate segment of this population in low-wage urban labor markets. These trends have far-reaching implications for the crisis of urban community health.

Much of the new migration to New York City comes from Latin America, the Caribbean, and Asia, with most new immigrants settling in the urban core, in the very neighborhoods plagued by poor housing conditions and concentrations of toxic hazards. The patterns in the other major U.S. poles of immigration are similar. This chapter is concerned with the nexus of immigration, race, class, and environmental risks to health in an urban neighborhood. It asks: as immigrants and their children make their lives in the congested, polluted centers of the urban landscape, how can public health officials help them avoid childhood diseases like lead poisoning and asthma, associated with living in low-income urban neighborhoods?

For urban health professionals, the juncture of expanded cultural diversity, poor environmental quality, and persistent poverty presents unique challenges to preventing and treating a host of illnesses that continue to plague the urban poor. Many in the public health field are approaching solutions to these problems with a renewed emphasis on family, social networks, and communal resources. This is similar to the relatively recent turn in the social sciences to the concept of social capital to explain a wide variety of social and political phenomena. Social capital is commonly defined as the resources that inhere in relationships of trust and cooperation among people. As a concept about differentially distributed but not fixed socially held assets, it has been used in an expanding range of applications, especially in the applied policy fields and in studies on poverty. But social capital is also a contested concept generating much controversy over its meaning, application, and effects, and some scholars have called for a more careful specification of how it is created and acquired, and what it can be used to explain. Portes, for example, questions the use of social capital explanations while the theory remains underdeveloped (Portes 2000). Despite the theoretical problems, the concept of social capital is appealing to many public health researchers because it is an integrative concept that can capture something important about the complex, community-level social and economic processes influencing public health, depending on how the concept is operationalized in empirical research designs.

The Sunset Park Health Study

As an emerging research paradigm in the field of public health, studies that empirically examine the relationship between social capital and public health are few (James, Schulz, and van Olphen 2001). We report on one such study designed to collect data on attitudes toward health care, child health issues, and awareness of environmental hazards, in particular by low-income parents living in Sunset Park, Brooklyn. The purpose of our study is to help researchers and health care providers address the health information needs of immigrants living in low-income urban neighborhoods plagued by environmental hazards.

The major research questions of the study were: Among low-income residents of an urban neighborhood, what is the relationship between the physical environment (including housing conditions) and the health risks it poses? What do low-income parents know about childhood health problems that plague low-income urban communities, like asthma and lead-paint poisoning? What are the most effective means of educating low-income parents about childhood health risks, preventive medicine, and treatment options?

Methodology

To assess the level of knowledge about child environmental health risks among community residents, we conducted 262 interviews with randomly selected parents utilizing public parks in the low-income neighborhoods of Sunset Park, Brooklyn, on several late summer days in 1998.[1] The 102-question survey instrument was developed in consultation with representatives from Lutheran Medical Center, the largest provider of health services in the area and a community partner in the research project. The survey took approximately 45 minutes to administer and the instrument was translated into Spanish. The use of bilingual interviewers permitted us to conduct interviews easily with non-English-speaking Latino parents. As a result, about half of the respondents in the sample completed the survey in Spanish. Latinos were the largest ethnic group in the sample (83 percent). Most of those identifying as Latinos originated from the Spanish-speaking Caribbean (30 percent are Puerto Rican and 8 percent are Dominican, with another 27 percent Mexican, and 14 percent claiming Central American birth or ancestry). As this is a largely new-immigrant sample, 42 percent said they spoke Spanish at home, while another 36 percent said they spoke English and Spanish at home.

Demographic Profile of Sunset Park, Brooklyn

Sunset Park is a mixed industrial, residential, and commercial area of about 120,000 people in southwest Brooklyn. As a large residential neighborhood it is distinctive for its industrial zone, encompassing the entire waterfront from Bay Ridge on the south to Red Hook on the north. Over the course of the last century, the waterfront economy shaped Sunset Park's development as a multi-ethnic, and later, multi-racial working-class community. With the demise of the shipping industry in New York Harbor in the 1950s, Sunset Park entered into a three-decade period of decline from which it has only recently begun to emerge. A study of Sunset Park in the 1980s analyzed the impact of the new immigration from Asia and Latin America on the area and found positive effects of demographic renewal in tempering the impact of de-industrialization (Winnick 1990). According to the study's author, immigrants to Sunset Park have reversed population losses and revitalized local commerce. Combined with an aging and declining white population, immigration has made Sunset Park a significant "minority-majority" residential area with a distinctive Latino influence (a majority of the population come from the Hispanic Caribbean, Mexico, and Central America). In addition, the Asian population in Sunset Park has increased by more than 100 percent in the last two decades, creating the city's third major Chinatown. With its excellent access to public transportation and the low-rise character of its housing stock, the area has been a destination of choice for striving, entrepreneurial newcomers.

And yet, many problems remain, including poverty, substandard housing, overcrowded schools, and environmental health risks associated with industry and the Gowanus Expressway, a congested elevated highway built through the area in the 1940s (Katinas 1998). Accordingly, nearly the entire neighborhood, from the waterfront eastward through its residential core, has been designated a Brownfield Tax Incentive Zone by federal and state environmental agencies engaged in promoting the cleanup, redevelopment, and reuse of contaminated properties in distressed urban communities (U.S. Environmental Protection Agency 2003).

The 1996 New York City Housing and Vacancy Survey (HVS)[2] reports that the median household income in the Sunset Park sub-borough area was $28,500, just below the city median of $29,600.[3] Lower income means higher rates of poverty and public assistance. The poverty rate in Sunset Park in 1996 was 23.7 percent of households, with 24.0 percent of households receiving public assistance, compared to a 20.6 percent poverty rate citywide, with 19.2 percent of households across the city receiving public assistance. Nearly half (47 percent) of all adult residents in Sunset Park have less than twelve years of schooling, compared to one third of the adult population citywide. Two in five residents of the area are age 25 or younger, and another nine percent are over 65, creating a substantial mostly non-working population of special needs (New York City, Department of City Planning 2003).

Results

Health education efforts among education and community-based health professionals we consulted for the survey identified low-income immigrant parents as a resident population most in need of child health education efforts.[4] The Sunset Park Health Survey sample, therefore, represents the low-income, young, female, Hispanic Caribbean parent population of Sunset Park (see Table 1). Half of the respondents in the survey report family incomes of less than $20,000 a year; all together, 86 percent of respondents report family incomes of $40,000 or less. Nearly half (46 percent) report that their children participate

TABLE 1   *Sunset Park Health Survey*
Socioeconomic Indicators

Gender	
female	76.5
male	23.5
	(260)
Race	
white	15.7
black	.4
latino	82.6
other	1.6
	(248)
Age	
16–24	12.3
25–35	52.5
36–45	26.1
over 45	8.6
	(244)
Educational Attainment	
less than high school	37.4
high school diploma	37.9
some college or associate's degree	8.41
bachelor's degree	11.1
graduate/professional degree	5.3
	(190)
Family Income	
less than $12,000	20.4
$12,001–$20,000	30.8
$20,001–$30,000	20.9
$30,001–$40,000	13.9
over $40,000	14.0
	(201)

in school lunch programs, and almost one in five respondents currently receive public assistance (see Table 2). Rates of educational attainment are low, with three-quarters of the respondents holding a high school diploma or less. Screening for parents yielded a sample that is disproportionately female (77 percent); two-thirds of the respondents in the sample are 35 years of age or less. The majority are married (55 percent) or previously married, with only 11 percent reporting they are single and never married. The mean number of children is two.

Housing and Urban Environmental Conditions

New York City is unique among American cities for the relative size of its renter class. Seven in ten householders, or two million people in the city, rent their housing units. This ratio also applies to Sunset Park, where 74 percent of households rent their housing. Similarly, Sunset Park's housing quality mirrors

TABLE 2   *Sunset Park Health Survey*
Social Services

In Last Year, Received Help From:		
school breakfast or lunch program	45.5	(242)
child welfare/protective services	2.9	(240)
women's or family shelter	2.5	(241)
utility bill assistance	6.2	(241)
rent assistance	7.0	(242)
counseling services	7.1	(238)
legal services	7.6	(237)
job placement services	5.1	(236)
Currently Receives:		
social security	6.5	(231)
SSI	11.3	(231)
medicare	7.8	(232)
medicaid	38.3	(234)
unemployment insurance	7.4	(230)
public assistance	18.4	(234)
food stamps	18.8	(234)

the city's overall with respect to what the HVS calls "maintenance deficiencies." Maintenance deficiencies are observed by HVS interviewers who conduct the survey inside a housing unit. There are seven categories of maintenance deficiencies: inadequate heating; heating breakdowns; cracks or holes in the walls, ceilings, or floors; non-intact plaster or paint; the presence of rodents; inoperative toilets; and water leakage from outside the unit (Lee 1999, 357). Overall housing quality citywide with respect to maintenance of condition in 1996 was good, although quality varies considerably with housing structure class, age of the building, and rent levels. On balance only 6 percent of renter-occupied units had five or more maintenance deficiencies, both in the city and in Sunset Park, in 1996.

The vast majority of respondents in the Sunset Park Health Survey, 86 percent, live in rental housing of very poor quality. Two in five report living in overcrowded conditions (more than 1 person per room); three in ten have broken plaster or peeling paint inside their apartments. Nearly two-thirds said that in the last three months they had seen mice or rats in their buildings, while three-quarters reported seeing cockroaches. Many respondents have problems with heat and grievously substandard building conditions. A quarter report that the heating systems in their apartments broke down during the previous winter, while nearly a third said that water had leaked into their apartments during the last year (see Table 4). Fifteen percent said that in the previous three months their toilets did not work for at least six consecutive hours. Most notably, one in ten said that their buildings lacked complete plumbing facilities altogether—that is, hot and cold piped water, a flush toilet, and a bathtub or shower.

TABLE 3   *Sunset Park Health Survey*
Ancestry, Language, and Religion

Ancestry	
Puerto Rico	29.4
Mexico	26.9
Central America	13.9
Dominican Republic	8.4
Europe	6.3
Poland	3.4
Russia/Ukraine	2.5
other	9.2
	(238)
Language Spoken at Home	
Spanish	41.9
English/Spanish	35.5
English	16.5
other	3.6
English/other	2.4
	(248)
Religious Preference	
Catholic	72.9
Protestant	15.0
other	2.0
none	10.0
	(240)
Church Attendance	
several times a week	8.0
every week	26.6
almost every week	17.3
once or twice a month	17.3
a few times a year	24.9
never	5.9
	(237)

Many respondents are aware of the industrial character and noxious conditions of their neighborhoods in Sunset Park. Forty-four percent said they live within ten blocks of a factory; over half live near an auto body shop or major highway. A quarter live near the Gowanus Canal, a putrid waterway and the site of industrial dumping for decades, while 12 percent said that they live near a sewage treatment plant. Nearly half live near buildings with boarded-up or broken windows.

Neighborhood Health Services

To establish a baseline of knowledge about health care and health issues, we queried respondents about their attitudes toward health care and their use of doctors and medical facilities. Despite the status of Sunset Park as a working-

TABLE 4   *Sunset Park Health Survey
Environment and Housing Quality*

Within Ten Blocks of Residence:		
factory	44.4	(259)
Gowanus Canal	22.4	(254)
dry cleaners with cleaning plant	69.3	(257)
major highway	51.8	(257)
sewage treatment plant	12.2	(254)
auto body shop	50.4	(257)
Housing Type		
apartment	59.2	
multi-family house	29.2	
single-family house	8.1	
condo/co-op	2.7	
other	0.8	
	(260)	
Housing Tenure		
rent	86.2	
own	11.9	
neither	1.5	
	(259)	
Housing Conditions		
broken plaster/peeling paint inside apartment	30.7	(251)
overcrowded	17.4	(247)
severely overcrowded	22.3	(247)
In Last 90 Days:		
saw mice or rats in building	60.0	(260)
saw cockroaches	75.9	(253)
broken toilet	14.7	(258)
In Last Winter:		
broken heating system	23.5	(255)
In Last Year:		
water leaked into apartment	29.7	(249)

class, immigrant neighborhood and their own low-income status, a majority of survey respondents (55 percent) feel that the health care services available to them are about the same as those available to people living in other neighborhoods in New York City. There is a high level (66 percent) of reported utilization of services at the principal health care institution in the area, Lutheran Medical Center, with a main hospital at 55th Street and four other smaller facilities nearby. Three quarters of those using Lutheran said that they went to the main facility at 55th Street to obtain care (see Table 5).

Access to health insurance is a determinant of access to health care. A

TABLE 5   *Sunset Park Health Survey*
Health Services and Insurance

Use Lutheran Medical Center Services	65.7
	(254)
LMC Location	
Sunset Park	74.4
Sunset Terrace	13.8
Park Slope	1.3
Family Physician	8.8
Park Ridge	1.9
	(160)
Access to Health Insurance	
Has Insurance	85.1
Does Not Have Insurance	11.0
	(228)
Type of Insurance	
Private	37.6
Medicaid	39.2
Child Health Plus	16.0
Other	6.2
	(194)
Where Family Goes for Medical Attention	
Private Neighborhood Doctor	28.3
Other Private Doctor	20.1
Walk-in Clinic (Not in Hospital)	15.2
Hospital Outpatient Clinic	21.3
Emergency Room	13.1
Other	2.0
	(244)

surprising 85 percent of respondents told us they had insurance of one kind or another. Among those reporting such, the most common form of insurance is that obtained through publicly financed programs such as Medicaid (39 percent), with another 16 percent using Child Health Plus, an insurance program open to uninsured school-age children in New York State. Two-thirds of respondents said that they had their own doctors, but this does not mean that the family practices preventive care or consistently uses a personal doctor when in need of medical assistance. Half of the respondents said that they use a walk-in clinic, emergency room, or other hospital clinic when in need of medical attention. The convenience of the location is the main reason respondents cited for going to these places for medical attention.

Levels of satisfaction with doctors are high. Eighty-four percent of respondents said that they were fairly or very satisfied with their last doctor's visit, and three-quarters had been to see a doctor in the previous year (see Table 6). Most respondents (72 percent) report excellent or good health. These findings

TABLE 6   *Sunset Park Health Survey*
Health and Doctors

Satisfaction with Last Doctor's Visit	
very satisfied	27.5
fairly satisfied	56.9
not too satisfied	9.0
not at all satisfied	1.2
no recent visit	4.7
	(255)
Quality of Health Over Last Few Years	
excellent	15.6
good	56.0
fair	24.9
poor	3.5
	(257)
When First Contacted Doctor About Problem	
at first sign of trouble	53.3
after problem has continued	33.3
wait until problem is serious	11.4
	(255)
Length of Time Since Last Doctor's Visit	
less than 6 months	35.2
6–9 months	26.2
9 months–1 year	14.8
1–2 years	12.9
more than 2 years	9.0
	(256)

are consistent with what we know about general trends in satisfaction with individual care—that is, people who are in good health and in regular contact with a doctor tend to report high levels of satisfaction with their own care. A large majority (80 percent) of respondents also report that they received good advice about preventive health care and healthy living from the medical staff with whom they interact. Respondents' behavior, however, appears to be at odds with this advice. Just under half of the respondents said that they first contact a doctor *after* a problem has continued or has become serious. A slim majority of respondents (53 percent) said that they or their families actually receive preventive health care, with cost and affordability as the major reasons given for not seeking such care.

Child's Health Status

Asthma is a chronic, inflammatory disease of the lungs. Attacks are triggered usually by allergens in the air and bring on constriction of the air pas-

TABLE 7   *Sunset Park Health Survey*
Child Health Issues

Number of Children		
1	35.6	
2	36.4	
3	20.0	
4	5.6	
5 or more	2.4	
	(250)	
Children Suffer From:		
asthma	20.0	(250)
allergies	14.9	(248)
upper respiratory infections	10.1	(247)
other breathing problems	12.7	(244)
ear infections	14.9	(248)
bronchiolitis	4.9	(247)
Worried About Child's Health Problem	23.9	(247)
asthma	47.2	
frequent colds	13.2	
speech impairment	9.4	
other problem	30.2	
	(53)	
Over Past Two Years:		
child's average hospitalizations	0.84	
child's average emergency room visits	1.25	

sages to the lungs, with coughing and wheezing. Morbidity and mortality from asthma are on the rise in the United States for reasons that elude medical researchers, although common theories point to increased exposure to indoor allergens and outdoor environmental irritants, changes in the immune systems of children, and poor access to good quality health care services.[5] Although asthma afflicts people of all ages and races, asthma rates in big cities are skyrocketing. Because asthma is associated with deficient housing quality and a lack of access to health care, it is increasingly a disease of the urban minority poor (Beckett et al. 1996; Steigman 1996; Stolberg 1999).

One-quarter of the Sunset Park Health Survey respondents expressed concerns about the health of their children, with asthma and allergies the top concerns. Asthma ranked first as an ailment among a group of common health problems in children, including upper respiratory infections, ear infections, bronchiolitis, allergies, and breathing problems other than asthma. Twenty percent of respondents said that their children suffered from asthma (see Table 7). One third of respondents' children had been hospitalized once or twice in the previous two years, with asthma as the most common reason for the hospitalization.

Forty-two percent of respondents said that their child had been seen in an emergency room once or twice over the previous two years. Again, asthma was a principal reason for the emergency room visit, second only to high fever.[6]

Indeed, a recent study found that children in New York City are almost three times as likely to be hospitalized for asthma as children nationally. In the decade following 1988, asthma hospitalization rates in New York City increased overall by 22 percent. However, the largest increases (upwards of 60 percent) were among children from low-income communities. These children were more than four times more likely to be hospitalized for asthma than were children from high-income areas (New York City Department of Health 1999; see also Claudio et al. 1999). Rates of asthma hospitalization among children fourteen years of age and younger increased in New York City over the last decade by 55 percent, from 9,275 (6.42 per 1,000 persons) in 1988 to 14,780 (9.94 per 1,000 persons) in 1997. The citywide rate is being driven by a 100 percent increase in the asthma hospitalization rate in the Bronx; hospitalizations increased by 39 percent in Brooklyn, but saw an inconsequential decrease in Sunset Park during this period, from 181 hospitalizations (7.69 per 1,000 persons) in 1988 to 180 (7.58 per 1,000 persons) in 1997.

Parents interviewed for the Sunset Park Health Survey demonstrated a high level of awareness of asthma and its symptoms. More than 90 percent said they knew what asthma was and were able to identify the symptoms of asthma as shortness of breath and coughing and wheezing. More than 90 percent said that they believed house pets can make asthma worse in a person who suffers from it. About half said that they thought regular exercise was bad for asthmatics, while about three-quarters said that exposure to cockroaches worsened asthma.[7]

Lead poisoning is another risk facing low-income residents of aging urban housing. As with asthma awareness, more than 90 percent of survey respondents said they knew how one could be poisoned by lead paint, that is, by inhaling or ingesting lead paint chips. Three-quarters of the survey respondents said that they had had their children tested for lead poisoning. Forty percent said they thought their child should be tested for lead now. Only one person in the sample said her child suffered from high levels of lead in the blood.

Discussion

The purpose of the Sunset Park Health Survey project was to ascertain the level of awareness about lead exposure and asthma among vulnerable community residents in Sunset Park in order to assist and inform future educational and outreach activities aimed at increasing preventive health behavior. The study suggests that the general level of awareness of these child health risks is high across a low-income immigrant population in Sunset Park where the expectation is that due to low levels of education and income the level of awareness would be low. Approximately 15 percent of the sample said they did not know about either asthma or the dangers of lead-paint poisoning. We compared this

TABLE 8   *Sunset Park Health Survey*
"Community Connectedness" Measures

	Awareness of Child Health Risk		
	Low	High	
Attend Church Weekly or More	21.2	36.9	(231)
Registered to Vote	14.7	43.7	(240)
Vote in School Board Elections	15.4	30.7	(231)
Member of a PTA	10.7	20.8	(230)

low-knowledge group with those who said they knew about the risks of both conditions to test factors that could be related to lower levels of awareness of child health risks. Tables 8 and 9 report the results.

We hypothesized that awareness increased in two major ways: first, through measures of what we call "community connectedness" that highlight a level of community involvement with social institutions like schools and religious and political organizations where community residents can receive information and educate themselves about issues of concern; and second, through experiences with hospitalization and the health care system in general, where parents have opportunities to address the health issues of their children with health professionals.

Social capital theorists like James Coleman and Robert Putnam have revived the idea that bonds of community built on mutual trust and respect, and (especially for Putnam) through cooperation and common cause in acting together, create "resources" that become available to community members to help them solve their social problems and meet their social needs (Coleman 1988, 1993a, 1993b; Putnam 1995, 2000). We were particularly interested in a form of social capital that could be created through respondents' affiliation with or connection to community-based institutions like churches and schools that do not have high costs associated with membership and serve as centers of community life where information about issues of concern to the community can be disseminated. Since the education of parents about childhood disease is important for its prevention and treatment, we hypothesized that low-income parents would benefit from the ties they formed with community-based institutions like churches and schools. To measure this kind of social capital we solicited information from our survey respondents about their patterns of church attendance and whether they were involved in their children's schools through Parent-Teacher Associations. We also asked questions about voter registration and voting in school board elections to gauge parents' involvement in the schools.

In the comparison between parents with low and high levels of awareness there are indeed provocative differences in reported behavior with respect to the "community connectedness" measures we devised (see Table 8). For example, parents who are less knowledgeable about asthma and lead-paint poisoning are also less likely to frequently (every week or more) attend church (21 percent of "low awareness" parents report this behavior, compared to 37 percent of the "high

TABLE 9   *Sunset Park Health Survey*
Child Health, Health Services and Insurance

	Awareness of Child Health Rise		
	Low	*High*	
Missed Days of School Last Year			
none	26.3	17.3	(129)
1–5	63.3	63.6	(129)
6–10	0.0	9.9	(129)
More than 10	0.0	4.5	(129)
In Last Year:			
Frequency of Child's Visits to Doctor's Office			
none	17.4	2.9	(161)
1–2 visits	52.2	41.3	(161)
more than 2 visits	26.0	49.8	(161)
hospital(izations)			
none	71.4	58.0	(235)
1–2 visits	25.8	34.5	(235)
more than 2 visits	2.9	7.0	(235)
emergency room			
none	62.9	36.8	(228)
1–2 visits	31.4	47.7	(228)
more than 2 visits	2.9	14.9	(228)
Use Lutheran Medical Center	41.7	70.0	(246)
Where Family Goes for Medical Attention			
private neighborhood doctor	36.4	26.2	(235)
other private doctor	21.2	19.8	(235)
walk-in clinic (not in hospital)	6.1	15.8	(235)
hospital out-patient clinic	15.2	23.3	(235)
emergency room	18.2	12.9	(235)
other	3.0	2.0	(235)
Health Insurance			
private	33.3	31.6	(220)
medicaid	45.5	31.6	(220)
child health plus	6.1	15.5	(220)
other	3.0	6.4	(220)

awareness" group). They are significantly less likely to say they are registered to vote (15 percent compared to 44 percent of the high awareness group), and half as likely to report voting in school board elections (15 percent compared to 31 percent of the high awareness parents). They are also half as likely to say they belong to their school's parent-teacher association (PTA).

Their lack of knowledge about asthma and lead-paint poisoning also reflects different patterns of child illness and utilization of the health care system. Deceptively, the children of these parents appear to be less sick in general than the children of parents demonstrating greater knowledge of asthma and lead paint poisoning (see Table 9). For example, 63 percent of the children of the less knowledgeable parents missed one or more days of school due to sickness over the previous year, compared to 78 percent of other children. Similarly, the former group were less likely to have seen a doctor, to have been hospitalized, or to have visited an emergency room in the previous two years than the children of the latter group. While less interaction with the health care system may mean that children are healthier, it also may not. It does suggest, however, that parents with less knowledge have fewer opportunities to learn about child health risks and disease.

In fact, 70 percent of the parents in our sample with the most knowledge of these issues report that they use the health care services of Lutheran Medical Center. Just over half that rate, or 42 percent, of least knowledgeable parents report using that facility. Less-informed parents are more likely to go to a private or neighborhood doctor for family care, whereas the better-informed parents are more likely to use walk-in or hospital clinics. The less knowledgeable parents tend to use emergency rooms more, perhaps because their children are less sick or they are not familiar with the range of available non-emergency room health care services. Patterns of utilization of health insurance suggest another reason: nearly half of the less knowledgeable parents reported that they use Medicaid to pay for their health care expenses, compared to a third of other parents. Both groups reported similar rates of private insurance (33 percent for less knowledgeable parents and 32 percent for others). The better-informed parents, however, are more than twice as likely (16 percent compared to 6 percent) to report their children enrolled in Child Health Plus, suggesting they are more "plugged in" to the options for care available to low-income parents.

The weaker associations and evidence of "community connectedness" for parents least aware of the risks to their children of asthma and lead-paint poisoning may or may not explain their lower levels of knowledge of these issues. This research is not so much aimed at finding the causes of lower levels of awareness risks as at suggesting directions for public health education among low-income, largely Hispanic Caribbean immigrant residents of an urban neighborhood.

Conclusion

Promoting health and preventing disease are critical functions of a viable public health system. As migration to the United States increases the nation's cultural diversity, it complicates the work of public health officials who must develop effective and efficient means of reaching new immigrants with information about prevention and treatment. Research has suggested that cultural practices, familial behavior, and communal norms may be important in understanding why some immigrant groups fare better or worse than expected according to traditional biomedical models of morbidity (Pearce 1996; Mendoza and Fuentes-Afflick 1999; Yen and Syme 1999; Ledogar et al. 2000). Our research in a Caribbean Hispanic immigrant neighborhood in New York City draws on these insights. It applies a social capital approach to understanding levels of awareness among low-income immigrant parents about their children's health and urban environmental health risks stemming from substandard and overcrowded housing, industrial pollution, auto emissions, and neighborhood decay. While many new immigrants receive better health care in the United States than they may have received in their home countries, issues of access continue to plague American urban health care delivery systems.

Our findings document a surprisingly high level of awareness among low-income immigrant parents of the environmental triggers of childhood diseases like asthma and lead poisoning, and suggest the awareness comes from dealing with higher levels of poor health and, not insignificantly, greater connection to community institutions like schools and churches where health promotion can be facilitated. Furthermore, our research supports an emerging family-community paradigm in the public health approach to environmentally induced health risks facing the urban minority and immigrant poor. The underlying logic of what Mendoza and Fuentes-Afflick call a "family-community health promotion model" emphasizes the strengths of these communities and reinforces their community connectedness as protective of optimum health behavior among community members.

The social capital of Hispanic Caribbean immigrant communities living in polluted and economically distressed urban neighborhoods enhances the level of knowledge in the community about childhood health risks stemming from environmental hazards. It is an important foundation of health that ties the individual's health to that of the community. As such, health care providers need to understand better and build on the communal and cultural practices that facilitate health promotion among low-income minority and immigrant communities. Our research suggests this as a promising approach to helping community members better protect themselves against the hazards of a de-industrial urban environment.

Notes

1. We decided on a public space methodology rather than on a door-to-door canvass for eligible respondents for two major reasons. The first reason was cost. The original design called for door-to-door interviews to be conducted by trained student interviewers. Accurately done, a door-to-door methodology for even a small sample is costly in both resources and time. Moreover, costs are inflated for door-to-door surveys in densely populated urban neighborhoods where people live in apartment buildings, have high mobility rates, and legal residence is often not easily established. An appropriate design requires that a sample be drawn from an accurate enumeration of all households in the study area. Because of our interest in reaching the lowest-income

residents in Sunset Park, many of whom are likely to be undocumented immigrants living in doubled-up housing situations, we felt that we could not rely on the 1990 census enumeration and that an original enumeration needed to be made first before we could draw a sample. Even if we had had the resources to conduct an original enumeration, we had other concerns with the door-to-door methodology: namely, responsibility for the safety of our student interviewers, and the difficulties of supervising field work that needed to take place in the off-hours of the day when people might be home. We therefore decided to conduct the survey in public gathering places where we might encounter low-income and immigrant residents. This facilitated supervision of the fieldwork, and we felt that by personally approaching people during their leisure we had a better chance of persuading them, especially undocumented immigrants, to participate in the survey project.

2. The New York City Housing and Vacancy Survey (HVS), conducted for the City at least once every three years by the U.S. Census Bureau in compliance with New York state and city rent regulation laws, is the best source of reliable inter-decennial census data. The large sample size (approximately 18,000 households) permits disaggregation of the data to 54 "sub-borough" areas that closely approximate the City's 59 community board boundaries. In the 1996 HVS the sub-borough area including Sunset Park varies little from Community Board 7's boundaries. Both Community Board 7 and the HVS Sunset Park sub-borough area include a smaller neighborhood to Sunset Park's north and east called Windsor Terrace, which comprises approximately 12 to 15 additional census tracts (Lee 1999).

3. Mean household size was 2.6 persons.

4. The survey project was funded by Brooklyn College through a grant from the U.S. Department of Housing and Urban Development, and developed through consultation with officials from Lutheran Medical Center.

5. According to the Centers for Disease Control and Prevention, the number of people suffering from asthma in the United States has increased over the last two decades from 6.7 million to an estimated 17.3 million people in 1998 (Vogel 1997).

6. According to the New York City Department of Health Neighborhood Health Profiles, after pneumonia and influenza, and injuries and poisonings, asthma is the third leading cause for hospitalization among children under the age of nine in Sunset Park (New York City Department of Health 2003).

7. Recent research suggests the carcasses of cockroaches produce a fine dust as they decay that in some people can trigger an asthma attack (Rosenstreich et al. 1997).

Conclusion

TOWARD A CREOLE ENVIRONMENTALISM

BARBARA DEUTSCH LYNCH

*T*he images of sun, sand, and sea that fuel the Caribbean tourist economy conceal more than they reveal about island environments. The environmental and societal side effects of these corporate constructions of secular paradise are carefully airbrushed away. As Sheller (2003, 64) observes, "It is the editing out of things that do not fit which enables this fantasy 'torrid zone' to be unceasingly packaged and sold for Northern consumers." These consumers, the tourists who flock to the region's beach resorts, seldom see the grittier activities that scar the islands of their dreams—military maneuvers, mining, manufacturing, toxic agriculture, and urban sprawl. Nor do they see the open dumps or sea floors where their wastes end up. It is these realities, which Caribbean peoples confront on a daily basis, that the essays in this volume address. The essays in this volume draw sober, at times pessimistic conclusions about the potential for positive environmental change. However, at a deeper level they are somewhat more optimistic: they suggest the potential for a rooted, creole environmentalism that is not only in touch with Caribbean realities, but capable of transforming them.

We opened this volume with a reference to the quest for the holy grail of sustainability. If we define sustainability as meaning that if we keep on doing what we're doing, we can keep on doing it, then the prospects for sustainability in the island Caribbean do not look good.[1] Where land masses are small, as they are in the island Caribbean, competition for land can be fierce. Both the region's ecosystems and its spaces for cultural reproduction are being threatened as landscapes are commodified and consumed by tourism, agribusiness, mining, export platform industries, and real estate development. Struggle over these spaces for cultural and social reproduction often appear as environmental struggles, and therefore as struggles over the definition of environmentalism. Conversely, environmental agendas are frequently parts of broader social and political projects. Therefore, we need to ask not only whether the islands' economies and ecologies will be sustainable over the long term, but also what kinds of social and political arrangements a given definition of sustainability implies.

Where spaces for cultural reproduction are diminished to the point that sustainability is no longer possible, Caribbean people move elsewhere. As enriching as this has been for receiving countries, migration, like environmental degradation, can be a symptom of a pervasive regional malaise caused by inequitable and dependent development.[2] In presenting subaltern perspectives on the etiology of environmental degradation, contributors to this volume offer some new tools for understanding the contexts in which Caribbean environmentalism has arisen, the issues of concern to island residents, and the ways in which environmental movements have evolved to address these issues. A careful reading of these essays also allows us to identify elements in an agenda for environmental action.

The Context

Since—perhaps even before—Columbus's landfall at Punta Isabella, the environmental fate of the Caribbean has been intimately connected to the political economy of the world system and to the political ambitions of successive colonial and imperial powers—Spain, Great Britain, France, the Netherlands, and the United States. This connection has been a major factor in shaping and constraining environmental movements, programs, and policies in the region. A second contextual factor noted by several contributors to this volume is the diversity of the region's physical landscapes, ethnic and linguistic groups, and governing bodies. While this diversity is responsible for the remarkable cultural productivity of the region, it can impede concerted environmental action. The third contextual factor worth noting is the region's ecological fragility and its vulnerability to natural as well as human-induced disasters.

Structural Context

A weakness of the environmental movement based in the global North has been its emphasis on agency and its relative neglect of the structures of economic and political control. We ignore such structures at our peril. Since 1492, decisions and policies made elsewhere have had enormous environmental and economic consequences for the Caribbean. Sidney Mintz (1985) provides ample evidence of the socially and environmentally transformative role of sugar production to feed the industrial revolution in Europe and America, and the literary critic Raymond Williams (1973) notes how the gentle lifestyle of Jane Austen's heroines was supported, albeit invisibly, by the brutal economy of the Caribbean sugar plantation.[3] Even after the abolition of slavery, plantation agriculture continued to define the region as a massive labor camp.[4]

Today the structures of dependency take different forms. For example, the chapters by García-Martínez et al., Miller, and Burac indicate that transnational capital flows to core economies without creating the backward linkages that would propel endogenous economic development. The negative environmental consequences of these flows are not offset by the myriad environmental programs undertaken by bilateral and international assistance agencies. Miller notes the preponderant power of international capital vis-à-vis small Caribbean states and the impact of this imbalance on tourist investment. She argues that because the tourist industry is designed to facilitate the expatriation of wealth, it is all the more difficult for the Jamaican government to capture the resources needed to address the nation's growing environmental problems.

Caribbean colonial wealth was created by migrants to the islands—whether as slave, contract, or free labor. Migration remains a noteworthy aspect of dependent development in the Caribbean. On the one hand we find rural people leaving the countryside in search of jobs in export platform industry, or expelled from lands that have passed into the hands of tourism and real estate developers, agribusiness enterprises, or the state. Flight in the face of natural disaster remains a feature common to the region. Because the economies of the islands are small relative to the size of their populations, and because the backward and forward linkages on the islands are slender, migration to Venezuela, Colombia, and the cities of North America and Europe is common. As Soto-Lopez and Minnite and Ness note in this volume, migrants often end up in the decaying cities of the northeast, with their brownfields, aging infrastructure, waste transfer stations, and hazardous worksites. But this is not the whole story. Migrant income can provide island families with remittances that take the place of resources that are no longer available to them. It may also finance development on the islands and contribute to urban sprawl as it is invested in commercial development, condos, and showy retirement homes. Politically, migration has created new spaces for the sharing of information and the creation of transnational networks.

A third aspect of dependent development that merits discussion is the role of the international development community in shaping island environmental agendas. The policy objectives of international institutions, bilateral assistance agencies, and international NGOs do not always reflect Caribbean realities or the interests of Caribbean people. However, these institutions carry a great deal of weight because they fund the environmental programs of local governments and organizations (Paniagua Pascual 1998). To a significant degree, the kind of environmental research that gets done and the kind of science that gets taught reflects the agendas of these international agencies.[5] As a corollary, international development discourse is adapted by local institutions seeking funds.

Diversity

Cuban-American literary scholar Antonio Benítez-Rojo (1992) sees the island Caribbean as an infinite array of islands (in both the literal and the figurative sense) that appear to be replicas of one another. Yet they are subtly different, and these differences account for the region's remarkable diversity—a diversity that takes a number of forms. First of all, few can agree on where the region begins and ends. The greater Caribbean can be seen as including not only the Greater and Lesser Antilles, but also the nations on the Caribbean coast of South and Central America, Mexico, and the Gulf states of the United States. Do we include the sea islands of Georgia or other black Atlantic communities in our definition? Do we include diaspora communities in New York, Hartford, Miami, London, and Madrid? We can demarcate the region as we will, but the way in which we define its problems and its solutions will depend upon where we draw its boundaries. This fluidity poses challenges, but at the same time creates opportunities for Caribbean environmental problem-solving.

If the island Caribbean's history of colonial conflict and imperialist adventure has left a legacy of continuing dependency, it has also given the region an extraordinary ethnic and cultural richness. An island like Cuba can boast of a population that is not only Taino, Afro-Cuban, and Iberian, but Mayan, Haitian, Jamaican, Levantine, Arab, Chinese, and eastern European. South Asians constitute a significant and active segment of eastern Caribbean populations. Migrants and conquerors alike have brought diverse political traditions to the islands; these have given rise to identity politics throughout the region. While identity politics are not a central focus of this volume, they comprise a leitmotif in many essays here. Can the search for identity give rise to a creole environmentalism?

The anthropologist Viranjini Munasinghe (2001) distinguishes two versions of the creole: the "callaloo" and the "tossed salad." The former implies a harmonious, well-cooked blend of cultures and traditions, while the latter suggests discrete cultural chunks, but in both cases the whole is greater than the sum of its parts. Her tossed salad metaphor is in many ways an apt description of Caribbean environmentalism. Rather than simply bemoan the fragmentation of the region's environmental movement, we might ask how disparate elements of the movement can be brought together to create a creole whole.

This task is made difficult by linguistic and political diversity. No pan-Caribbean discussion can take place without provision for translating at least into the major colonial languages—Spanish, English, and French. To the extent that environmental groups are part of a broader creole movement, it would make sense to provide for creole languages as well. This all makes coordination of environmental activity that much more difficult.[6]

The sheer number of states in the region is another impediment to environmental action. Significantly, there is no political organization that encompasses the entire island Caribbean. CARICOM, for example, includes neither Cuba nor the Dominican Republic, nor does it include Puerto Rico or Martinique. While the status of these latter islands is not quite comparable, what is noteworthy in both cases is the existence of a body of environmental law that is differentially implemented and enforced in the mother country and in the colony. As Valdés Pizzini and Burac show, this differential creates interesting political spaces where environmentalism has been coupled with demands for greater autonomy. Not surprisingly, the environmentalism we see in these contexts is not conservative or conservationist, but entails a radical critique of environmentally questionable development programs and policies. The Vieques movement is an excellent example of the latter. Environmentalism in the colonial context may well conform to what Ramón Grosfoguel (2003, 72) proposes as a new radical project focused on "the democratization of political power over the environment."

A critique of imperialism exists in the independent states of the region as well, although it is more ambiguous in the independent nations of Cuba and the Dominican Republic, both of which have sought to control the terms of their integration into the global economy. Both nations have eagerly sought foreign investment in mining and metallurgy, tourism, and export platform manufacturing as a way of generating the foreign exchange for oil and food imports. The environmental impacts of these revenue-generating activities have received somewhat less scrutiny than they have in Puerto Rico or Martinique. Rather, environmental activity has emphasized economic self-sufficiency (as in the case of urban agriculture and energy conservation programs in Cuba or self-help urban service delivery programs in the Dominican Republic) or tourism-linked conservation (as in the case of projects carried out by groups like the Center for Conservation and Eco-development of Samana Bay and its Surroundings [CEBSE]).

Fragility

Caribbean ecologies are fragile and highly vulnerable to natural as well as man-made disasters. Devastating hurricanes are increasingly frequent. Although they do greater damage to infrastructure than to natural systems, the flooding and landslides that accompany tropical storms can cause catastrophic landscape transformations even in natural areas like Puerto Rico's Luquilb National Forest. Island scientists, policy makers, and beachfront developers fear the impacts of global warming on storm frequency and sea level, but not all perils are storm-induced. The higher islands are also subject to earthquakes and volcanic eruptions; the latter have rendered entire islands unfit for habitation.

As ecosystems, the islands have high rates of endemism, but relatively low levels of biodiversity. The barrier reefs that surround the low islands are being destroyed by land-based pollution and toxic runoff. Fish populations are endangered by reef and mangrove destruction as well as by ocean dumping, oil drilling, sedimentation and pesticide runoff, and overfishing both inshore and on the open seas.

Sugar cultivation and ill-conceived large-scale irrigation projects have taken their toll on the islands' ecosystems. As Lynch's chapter shows, traditional polycultural food production systems, pushed to marginal lands, are less able than ever to meet local food needs. Food security has thus become a serious concern for many island residents. A related concern raised by García-Martínez et al. in this volume—the safety and adequacy of drinking water—is a problem not just in Puerto Rico, but in most of the Antilles.

In sum, the island Caribbean suffers from vulnerability to global economic and environmental phenomena ranging from capitalism to climate change. This vulnerability is exacerbated by the small size of islands, by the relative powerlessness of Caribbean states, and by necessary dependence upon trade for economic survival. Out of this vulnerability arises a need for regional cooperation, and regional cooperation is growing stronger and more effective. The region faces a set of common environmental problems that can be addressed through concerted action that embraces the entire region while respecting its linguistic, political, and ethnic diversity.

The Issues

As Valdés Pizzini reminds us in his discussion of environmental movements in Puerto Rico's coastal zone, the environment is a field of action—an arena where civil society challenges the state. Contention within this field takes place not only between civil society and the state, but between civil society and transnational capital, and among groups in civil society. Political mobilization clusters around particular problems or issues—some more than others. Depending upon the issue, the type and intensity of political activity will vary. That said, what are the issues? The tourists who descend upon the islands in January or who ply the Caribbean in small craft, yachts, and floating hotels are likely to focus on the degradation of once pristine beaches and coastal waters. More adventurous eco-tourists seek national parks and protected areas with well-marked trails and folkloric displays, but otherwise free of human activity. International NGOs like the Nature Conservancy and R.A.R.E. have emphasized biodiversity protection and forest conservation. WWF, UNEP, and the Center for Marine Conservation have drawn our attention to marine pollution and coastal waters. If it is considered at all, the built environment is seen by many northern environmentalists as a theme park in which a sanitized narrative of conquest and colonization is enacted. Here, environmental concern is largely limited to the preservation of colonial architecture.

Not surprisingly, as our essays show, these issues are not always primary concerns for residents, whose immediate needs include secure access to land, shelter, and to the biotic resources they need to complement goods they can purchase with cash; a safe and adequate supply of water for drinking, bathing, and irrigating crops; air pollution control and safe handling of sewage and solid wastes. Given the immediacy of these needs, it is tempting to argue that environmental consciousness in the global South is nascent, at best.[7] But this would be to ignore the history of political mobilization in the region and the very real environmental concerns that have provoked it. Caribbean environmentalism tends to focus on brown rather than green issues; it addresses problems associated with urbanization, tourism, manufacturing and extractive industry on and around the islands.

García-Martínez et al. and Valdés Pizzini show that for Puerto Ricans, key issues are air and water pollution resulting from Operation Bootstrap industrialization. It can be said that Operation Bootstrap gave rise to the Puerto Rican environmental movement, which was at its outset a Left critique of industrial policies enacted on and on behalf of the island (Concepción 1995). The *Misión Industrial* is an exemplary case. In contrast, Dominican protest against industrial pollution generally consisted of isolated neighborhood campaigns lacking a broader vision.

Oil and mineral extraction has raised concern throughout the region. Hurricane Katrina offers Cuba a timely warning about the risks posed by the region's petrochemical industry as it joins Trinidad and Aruba as an oil-producing nation. In the Dominican Republic, NGOs and local residents have protested pollution and land degradation resulting from gold mining near Cotui. Yet as Andrés Serbin argues, Caribbean Basin environmental degradation has "a clear supranational angle that includes chemical pollution produced by industry and agriculture, the dredging, filling in and poor use of land; the irrational exploitation of coastal and marine resources; and pollution produced from waste from coastal cities and inland centers." Absent still is a region-wide coordinated effort focused on these issues.

Urban sprawl is of growing concern in Puerto Rico (Valdés Pizzini, chapter 4 of this volume), Cuba, and in the Dominican Republic (Lynch, chapter 7 of this volume). As García-Martínez et al. (chapter 6) show, the operation of real estate markets ensures that the poor will live in areas most affected by pollution. Pollution problems are aggravated by the fact that the residents of the region's spreading cities rely on used cars, buses, and trucks fueled by cheap, dirty petroleum. Sprawl also complicates waste management, a task that often falls to poorly funded municipal governments. Where sprawl is a problem, municipal resources may not be sufficient to pay for new sewer lines and connections, for garbage hauling, or for the construction of sanitary landfills.

Sprawl rarely solves the problem of shelter, and those who work in low-wage urban jobs rely on housing in informal settlements or shanty towns. Often, as part of state and municipal environmental programs, informal settlements are eradicated when they lie in what are perceived as ecologically sensitive areas. Evictions mean a loss of physical spaces for social life. Coyula (1996) asks us to rethink the "barrio insalubre" or "foco de miseria." and to develop an appreciation for the capacity for change and adaptation within these communities— a capacity that is largely absent in the public housing sector.

A corollary of sprawl noted by Lynch is the withdrawal of land from food production, one of many threats to traditional Caribbean food systems. The emergence of urban agriculture is a stopgap response to sprawl and the loss of land for food production. A more encouraging trend is the shift to organic production, which began in Cuba during the Special Period. However, we can see a notable turnaround even in the Dominican Republic, which had been highly dependent on pesticide-intensive agriculture well into the 1990s. This turnaround is tied not just to the failure to eradicate *thrips palmi* and white fly, but also to growing markets in Europe and North America for organic products.

Access issues are fundamental for those who depend upon natural resources for their livelihoods. Environmental policies shaped with international assistance often overemphasize the creation of protected areas (see Lynch 2001). This is consistent with international insistence on the simplification of property rights. Valdés Pizzini shows that while Puerto Rican environmentalists have fought against huge tourist developments that would threaten the ecological integrity of coastal zones, they are equally opposed to protected-area projects that would restrict fishing and other community activities. The conflict here is between competing approaches to environmental protection: one favoring exclusion of resource users and another favoring prohibition or restriction of polluting or otherwise destructive land uses. Eviction for tourist development has been a persistent target of movement activists.

The public health consequences of environmental degradation are also important issues for Caribbean environmentalists. Dominican environmental groups have drawn attention to diseases associated with agrochemicals and contaminated water supplies (Lynch, chapter 4 of this volume; see also Lynch 2001). García-Martínez et al. eloquently describe the cycles of sickness, unemployment, and poverty that have plagued Puerto Rico, the cancer clusters in Cataño, and the respiratory problems suffered by those who live close to poorly regulated industrial sites. Minnite and Ness (chapter 10) and Soto Lopez (chapter 9) find that these concerns are also central to the environmental justice movement in the northeast, which owes much to the organizing efforts of Caribbean Latinos.

The region's environmental problems are often global in their origins, regional in their scope, and local in their manifestations. Ocean dumping by unregulated cruise ships and freighters is a good example: flying flags of convenience, these ships are subject to few controls and regulations. Tourism is another: large tourist enterprises often are far better capitalized than the national governments they seek to influence. Hence they are often in a position to determine the terms of trade, as Miller and Burac illustrate in their chapters on Jamaica and Martinique. Agricultural practices are by and large determined by large transnational agribusiness enterprises, whose presence in any given country is likely to be short term. Finally, as the forts that dot the islands attest, the Caribbean has traditionally had military importance for European and North American powers. As McCaffrey and Baver (chapter 8) argue, movements to force base closures in Puerto Rico reflect the fact that local environmental problems cannot be confronted without looking beyond national borders.

The Movement and Organizations

In 1992, three years before the CUNY workshop, I was told that Caribbean environmentalism needed a jump start. If the environmental movement did not exist, NGOs, foundations, universities, and aid agencies from the North could help bring it into being. At the time this comment surprised me, because it ignored what I saw as obvious signs of a lively movement and committed activism in the region. Upon reflection, however, I concluded that because Caribbean environmentalism did not conform to the expectations of its potential northern supporters, its issues and even its existence went unrecognized.

The Caribbean environmental movement reflects the region's diversity. Diversity invites typological exercises. Jácome and Valdés Pizzini do an excellent job of teasing apart the various tendencies within the movement as it has evolved in the past twenty-odd years.

Paradigms

One way to understand the diversity among movement organizations is to look at the paradigms that undergird their programs. It is convenient to divide environmental paradigms into three basic groups, each deriving from a different tendency within environmental thought: many members of conservationist and preservationist groups come from the biological sciences and/or subscribe to a neo-Malthusian paradigm; advocates of sustainable development often come from an ecological modernization perspective, while environmental justice activists ground their arguments in political ecology.

Conservationist organizations still abound, and these often find support from international NGOs. They often focus on protected-area management, and they tend to identify economic activities of poor rural residents—e.g., charcoal-making and shifting cultivation—as causes of degradation. The sustainable development paradigm—articulated in *Our Common Future* (WCED 1987)—has informed the activities of many developmentalist organizations funded by bilateral assistance programs and international agencies. As it has evolved in the past decade, the sustainable development approach has increasingly emphasized its roots in mainstream modernization theories (see Harvey 1996; Dryzek 1997; Lynch 2001) and has turned to the market and to industrialists for solutions to environmental ills. Within the ecological modernization paradigm, pollution and environmental health risks are seen as symptoms of backwardness that will disappear as countries advance. Programs based on this notion of a "risk transition" tend to address risks associated with water-borne bacterial disease, and to neglect risks directly associated with industrial, extractive, and military activity.

Jácome does not see significant mobilization around environmental justice on the islands, but it is easy to read the *Misión Industrial* in Puerto Rico as a prototypical environmental justice organization. In addition, the movement organizations that focus on evictions, health, and access described by Valdés Pizzini and Burac can be considered part of the broader environmental justice movement. The environmental justice paradigm in the North has also been shaped to a large extent by Caribbean Latinos, who bring to struggles in the North perspectives shaped by the islands. For this reason, if for no other, we can hope to see considerable cross-fertilization between environmental justice organizations in the South and in the North. The participation of residents of the U.S. mainland in the Vieques protests may be evidence of increasing cross-fertilization.

Institutional Forms

The institutional forms of Caribbean environmentalism also vary widely. NGOs are often seen as the main drivers of the movement, and those who study the movement tend to focus on NGO performance. This approach is at best limiting and at worst self-serving. Valdés Pizzini speaks to the role of the university as a focal point for environmental activity. One example is the Dominican CEBSE, which was formed by a cohort of biologists from the Autonomous University of Santo Domingo who specialized in marine and coastal resource management. Environmental activists can also be found in the ranks of government agencies, trade unions, and political parties. It is the presence of committed activists in various sectors that makes concerted environmental action possible. An example of such action is the program to clean up Havana's Río Almendares. The call to action came from an NGO, but it was eventually taken up by park authorities and local government bodies.

Looking at Caribbean environmental NGOs as institutions, we find a mixed bag, ranging from small, lean social movement organizations to well-heeled groups housed within government ministries. As Jácome notes, few NGOs are governed democratically, and most do not have mechanisms that hold them accountable to the groups that they purport to represent. Nonetheless, they are important to the functioning of the movement and have in many cases succeeded in linking their particular interests to issues of concern to large numbers of residents.

To understand Caribbean environmental institutions, we need to turn once again to the influence of Northern governmental and nongovernmental bodies on Caribbean environmentalism. According to Jâcome, "governments, environmental NGOs, and foundations from the Northern countries or international organizations in which funding from northern countries plays an important role provide most of the funding for environmental movements in the Caribbean" (chapter 2). She concludes that this will influence the goals and projects of both regional and local NGOs. Paniagua Pascual's (1998) research in the Dominican Republic suggests that at best this influence constrains Caribbean environmental programs, and at worst it can divert environmental-justice NGOs to less transformative pursuits. Another debilitating effect of this kind of influence is the preference of Northern organizations for dyadic, clientelistic relations. Although Jácome argues that international donors do little to foster horizontal linkages among Caribbean groups, we found a countervailing tendency of northern foundations to promote the development of dependent consortia and networks—sets of horizontal connections, but only within a larger vertical structure.

Linkages

At its best, Caribbean environmentalism is an environmentalism of everyday life. An important characteristic of regional environmental movements has been their ability to link environmental issues to other concerns. As elsewhere in the global South, a significant fraction of the Caribbean movement has abandoned a conservationist stance to engage in a broader critique of globalization and, as we noted in the Introduction, neoliberal policies and their impacts. While some argue that linking environmental to other social issues muddies the waters and makes it more difficult to achieve environmental goals, it may very well be that broadening the environmental agenda will broaden the movement's base.

Of note is the environmental movement's identification with struggles against colonialism and dependency. Puerto Rico's *Misión Industrial* is an example, as are the protests against the U.S. Navy presence on Vieques. In Martinique, as Burac notes in this volume, the ecology movement is contesting not only the consumption of natural spaces by the tourist industry, but the control of that industry by foreign interests. Finally, as Lynch shows, Cuba's shift to organic agriculture and its support for urban agriculture were motivated as much by a desire to reduce dependence on foreign inputs as by an interest in the environment.

Environment and public health are linked in a number of ways. Urban environmental activists who subscribe to the ecological modernization paradigm emphasize health problems associated with fecal contamination and insect-borne disease, while their counterparts in environmental justice organizations, in the unions, and in the women's movement focus on environmental health risks in the workplace and in the community.[8] Again, environmental justice activists are more likely to see public health issues both in the South and in the North in the context of economic globalization and dependent development.

A third important link ties environmental activism to issues of resource access and shelter. This linking of environment to claims of access to shelter and resources differentiates a number of Caribbean environmental organizations from their mainstream counterparts in the continental United States. It locates many of them squarely within an environmental justice framework. These movements bring to the environmental agenda such concerns as housing, public space, urban greening, and social life. Resource access is a key element in the environmental protests described by Valdés Pizzini, Burac, and Miller.

A weakness of early dependency theory was its failure to recognize linkages between the dependent development in the global South and distorted development within the hegemonic powers of the North. Soto-Lopez draws our attention to the intimate connection between U.S. programs like Operation Bootstrap and the Caribbean Basin Initiative—both of which had grave environmental consequences for the islands—and deindustrialization in the northeastern region of the United States, a process that has left brownfields and contaminated communities in its wake. The parallels between the environmental health problems encountered by Puerto Rican and Dominican migrants to the United States and Dominican rural migrants to the export processing zones are too striking to be ignored. These problems are encountered both in the community and in the workplace.

Toward a Creole Environmentalism

A major obstacle to concerted environmental action in the Caribbean is fragmentation of the movement, which in turn reflects a broader regional fragmentation and the diversity of interests in the environment. However, it also reflects a competition between local and transnational interests over control of the environmental agenda. International agencies tend to fund NGOs and programs that are technical-managerial in style and conservationist in tone, and that pose no threat to international capital in the region. If we look back to the issues that concern our authors and their colleagues, they have to do with controlling the behavior of international capital in the region—whether mining and petrochemical industries, agribusiness, or mass tourism.

Also implicit in many of the contributions to this volume is a redefinition of citizenship to include socio-environmental rights: the right to a healthy environment both in the community and in the workplace, the right of access to land and marine resources for food and shelter, and the political right to organize in defense of these rights. When authors talk about civil society and participation, it is within the context of this amplified definition of citizenship.

Many contributors to this volume call for increased participation in environmental decision making. Without reiterating the points on this topic raised in the introduction, I would like to underscore the distinction between participation and cooptation made by Soto-Lopez. He argues that when a community's participation is politically controlled, it is forced into the position of responding to outside initiatives rather than proposing its own. Jácome finds that this is a particularly grave problem for island environmental movement organizations that depend on international NGOs, foundations, and bilateral aid agencies for their operating budgets. It is also a serious problem for environmental NGOs that see process as an obstacle to solving immediate and severe environmental problems.

I would like to close not with prescriptions, but with a set of lessons that emerge both from the essays in this volume and from the experiences of activists and their constituents on the islands and in diasporic communities. The first is that there is great value in listening to Caribbean voices and giving them ample opportunities to define environmental agendas both in the North and in the South. Letting the diverse realities of Caribbean communities intrude into environmental planning processes can open them up to new visions and new ways of thinking. Second, the Caribbean environmental movement will grow stronger to the extent that local movements partake of North-South ties that emanate from the base rather than from the centers of power. And, finally, within Caribbean regions and neighborhoods, citizens need to search for new ways to exert influence on government agencies and create new fora for frank and open dialogues between industrial plant managers, agribusiness and tourism interests, workers, and neighborhood groups.

Notes

1. Thanks to John Nettleton of Cornell University Extension in New York City for sharing this definition.
2. Sheller (2003, 194) sees migration as integral to "creolization," and argues that this concept was initially theorized "not only in terms of mixture and mobility, but also in terms of conflict, trauma, rupture, and the violence of uprooting."
3. Stuart Hall (1991, 48) carries Williams's point a step further when he argues for the centrality of the Caribbean in English history: "I am the sweet tooth, the sugar plantations that rotted generations of English children's teeth . . . that is the outside history that is inside the history of the English. There is no English history without that other history."
4. I first heard the Caribbean described this way at a Cornell lecture by Jamaican scholar Rex Nettleford.
5. On this topic, see Michael Goldman (2001b).
6. Indeed, the Caribbean Natural Resources Institute (CANARI) staff have made the connection between creole language revival and environmental justice. This connection is also evident in Patrick Chamoiseau's (1997) novel *Texaco.*
7. This logic lies behind much international development thought (see for example the oft-cited Lawrence Summers memo of 1992).
8. See, for example, the work CIPAF, a Dominican women's NGO, did in the late 1990s on workplace environmental hazards in export-platform manufacturing zones and their impacts on women's health.

REFERENCES

Acosta, Tamara Zoé. 1995. *Señalamientos Críticos al Concepto de Apoderamiento: Revisión a la Luz de las Luchas Ambientales Guaniqueñas.* Master's Thesis, Department of Psychology, University of Puerto Rico, Río Piedras.

Agyeman, Julian, Robert D. Bullard, and Bob Evans. 2003. *Just Sustainabilities: Development in an Unequal World.* Cambridge, Mass.: MIT Press.

Altieri, Miguel. 1987. *Agroecology: The Scientific Basis of Alternative Agriculture.* Boulder, Colo.: Westview Press.

Altieri, Miguel A., and Susanna B. Hecht, eds. 1990. *Agroecology and Small Farm Development.* Boca Raton, Fla.: CRC Press.

Álvarez Ruíz, Migdalia. 1991. "Desafío ambiental en los complejos turísticos." *Diálogo,* Universidad de Puerto Rico, May, p. 41.

Álvarez Ruíz, Migdalia, and Manuel Valdés-Pizzini. 1990. "Guánica: Espejo socioecológico para el sur de Puerto Rico." *Acta Científica* 4, nos. 1–3.

Anazagasti, José. 2000. "El Carbón es Muerte: The Politics of Energy-Related Technologies in Puerto Rico." In *Chicana/o Latina/o Studies for the 21st Century: New Perspectives on Mentorship and Research,* edited by Marcos Pizarro. Pullman: Washington State University.

Ansine, Janice. 1992. "Negril Hotel Site Still Under Study." *Jamaica Weekly Gleaner,* March 2, p. 8.

Arizpe, Lourdes, and Margarita Velázquez. 1994. "The Social Dimensions of Population." In *Population and Environment: Rethinking the Debate,* edited by Lourdes Arizpe, M. Pricilla Stone, and David C. Major. Boulder, Colo.: Westview Press.

Asociación de Industriales de Puerto Rico (AIPR). 2004. *Informe AIPR Final.* Available in: http://egp.rrp.upr.edu/Investigacion/ERGEstadoSituacionIndustrial.htm.

ASSAUPAMAR. 1994. *Project de Schema d'Aménàgement Regional: Contribution de l'ASSAUPAMAR.* Lamentin, Martinique: ASSAUPAMAR Publishing.

Atkinson, Robert D., and Randolph H. Court. 1998. *The New Economy Index: Understanding America's Economic Transformation.* Washington, D.C.: Progressive Policy Institute, Technology, Innovation and New Economy Project.

Atlantic Division, Naval Facilities Engineering Command, 1st Quarter 2004. "Environmental Restoration News." www.vieques-navy-env.org/.

Autoridad de Desperdicios Sólidos. 2003. *Plan Estratégico de Manejo y Disposición de Desperdicios Sólidos*. Government of Puerto Rico (draft).

Ayala, César J. 2001. "From Sugar Plantations to Military Bases: The U.S. Navy's Expropriations in Vieques, Puerto Rico, 1941–45." *Centro Journal* (spring): 22–43.

———. 2003. "Recent Works on Vieques, Colonialism, and Fisherman." *Centro Journal* (spring): 212–225.

Bacon, Peter. 1987. "Wetlands for Culture and Heritage Tourism." *Caribbean Park and Protected Area Bulletin* 5, no. 1: 9.

Barker, David, and Duncan F. M. McGregor, eds. 1995. *Environment and Development in the Caribbean: Geographical Perspectives*. Kingston, Jamaica: University of the West Indies Press.

Barnet, Miguel. 1994. *Biography of a Runaway Slave*. Trans. E. Nick Hill. Willimantic, Conn.: Curbstone Press.

Barreto, Amílcar Antonio. 2002. *Vieques, the Navy, and Puerto Rican Politics*. Gainesville: University of Florida Press.

Baud, Michiel. 1987. "The Origins of Capitalist Agriculture in the Dominican Republic." *Latin American Research Review* 22: 135–53.

———. 1995. *Peasants and Tobacco in the Dominican Republic, 1870–1930*. Knoxville: University of Tennessee Press.

Baver, Sherrie. 1993. *The Political Economy of Colonialism: The State and Industrialization in Puerto Rico*. New York: Praeger.

Beck, Ulrich. 1992. *Risk Society: Towards a New Modernity*. Trans. Mark Ritter. London: Sage Publications.

Beckett, William S., Kathleen Belanger, Janneane F. Gent, Theodore R. Holford, and Brian P. Leaderer. 1996. "Asthma among Puerto Rican Hispanics: A Multi-Ethnic Comparison Study of Risk Factors." *American Journal of Respiratory and Critical Care Medicine* 154: 894–99.

Been, Vicki. 1993. "What's Fairness Got to Do With It? Environmental Justice and the Siting of Locally Undesirable Land Uses." *Cornell Law Review* 78: 1001ff.

Benedetti, María H. 2000. *Vieques: The Challenge for Peace*. Sea Grant in the Caribbean, University of Puerto Rico Sea Grant College Program, April-June.

Benitez-Rojo, Antonio. 1992. *The Repeating Island: the Caribbean and the Postmodern Perspective*. Durham, NC: Duke University Press.

Berman-Santana, Deborah. 1996. *Kicking Off the Bootstraps: Environment, Development, and Community Power in Puerto Rico*. Tucson: University of Arizona Press.

Besson, Jean, and Janet Momsen, eds. 1987. *Land and Development in the Caribbean*. London: Macmillan.

Bolay, Eberhard. 1997. *The Dominican Republic: A Country between Rain Forest and Desert: Contributions to the Ecology of a Caribbean Island*. Wekersheim, Germany: Margraf Verlag.

Brereton, Vera Ann, ed. 1993. *Proceedings of the Third Caribbean Conference on Ecotourism. St. Michael, Barbados:* Caribbean Tourism Organization.

Bryant, Bunyan. 1995. *Environmental Justice: Issues, Policies, and Solutions*. Washington, D.C.: Island Press.

Bullard, Robert D. 1994. *Dumping in Dixie: Race, Class, and Environmental Quality*. 2d edition. Boulder, Colo.: Westview Press.

Caraballo, María del Carmen. 1991. "Río Grande: Desplazamiento social comunitario en la costa." Sociology seminar paper on coastal gentrification. Department of Social Sciences, University of Puerto Rico, Mayagüez.

Carro-Figueroa, Viviana. 2002. "Agricultural Decline and Food Import Dependency in Puerto Rico: A Historical Perspective on the Outcomes of Postwar Farm and Food Policies." *Caribbean Studies* 30. no. 2: 77–107.

Carroll, C. Ronald, John H. Vandermeer, and Peter Rosset. 1990. *Agroecology.* New York: McGraw-Hill.

Catton, William R., and Riley E. Dunlap. 1979. "Environmental Sociology." In *Annual Review of Sociology*, vol. 5: 243–273.

Centner, Terence, Warren Kriesel, and Andrew Keeler. 1996. "Environmental Justice and Toxic Releases: Establishing Evidence of Discriminatory Effect Based on Race and Not Income." *Wisconsin Environmental Law Journal* 3 (summer): 119ff.

Cerame-Vivas, Máximo. 1994. *El Atropello Ambientalista: Extermino de la Sensatez.* San Juan: Editorial Librotex.

CEUR (Centro de Estudios Urbanos y Regionales). 1993. *Informe final de los resultados del proyecto uso de suelo y producción de alimentos en la República Dominicana.* Vol. 1. Santiago: Pontificia Universidad Católica Madre y Maestra.

Chamoiseau, Patrick. 1997. *Texaco.* New York: Pantheon.

Chantada, Amparo. 1992. "Los canjes de deuda por naturaleza. El caso dominicano." *Nueva Sociedad* [Caracas]122.

Chaparro, Ruperto. 1998. "Desinversión y desinterés: La situación en el manejo de las playas de Puerto Rico." Planteamientos sobre política pública. Programa de Colegio Sea Grant de la Universidad de Puerto Rico. Publicación Número UPRSGCP-G–74.

Chase, Anthony R. 1993. "Assessing and Addressing Problems Posed by Environmental Racism." *Rutgers Law Review* 45: 335ff.

Checker, Melissa. " 'Like Nixon Coming to China': Finding Common Ground in a Multi-Ethnic Coalition for Environmental Justice." *Anthropological Quarterly* 74, no. 3 (June 2001): 135–147.

CIIES. 1992. "Sustancias Tóxicas en Puerto Rico." *Boletín del Centro de Información, Investigación y Educación Social* (CIIES), December.

Claudio, Luz, Leon Tulton, John Doucette, and Philip J. Landrigan. 1999. "Socioeconomic Factors and Asthma Hospitalization Rates in New York City." *Journal of Asthma* 36, no. 4: 343–350.

Cole, Luke W., and Sheila R. Foster, eds. 2001. *From the Ground Up: Environmental Racism and the Rise of the Environmental Justice Movement.* New York: New York University Press.

Coleman, James S. 1988. "Social Capital in the Creation of Human Capital." *American Journal of Sociology* 94, Supplement: S95–121.

———. 1993a. "The Design of Organizations and the Right to Act." *Sociological Forum* 8: 527–546.

———. 1993b. "The Rational Reconstruction of Society." *American Sociological Review* 58 (February): 1–15.

Collin, Robert W., and Robin Morris Collin. 1997. "Urban Environmentalism and Race." In *Urban Planning and the African American Community: In the Shadows*, ed. June Manning Thomas and Marsha Ritzdorf. Thousand Oaks, Calif.: Sage.

Collinson, Helen. 1996. *Green Guerrillas: Environmental Conflicts and Initiatives in Latin America and the Caribbean.* London: Latin American Bureau.

Concepción, Carmen. 1995. "The Origins of Modern Environmental Activism in Puerto Rico in the 1960s." *International Journal of Urban and Regional Research* 19: 112–128.

Congressional Research Service. 2004. "Environmental Cleanup at Vieques Island and Culebra Island." August 4.

Conseil Regional de la Martinique. 1995. *Schema d'Aménagement Regional.* Fort-de France: Conseil Regional.

Conway, Dennis. 1993. "The New Tourism in the Caribbean: Reappraising Market Segmentation." In Dennis J. Gayle and Jonathan N. Goodrich, *Tourism Marketing and Management in the Caribbean*, 170–171. London: Routledge.

Corral, Julio. 2003. *La Gestión del Desarrollo Comunitario: Experiencia en los sectores Bayona y Honduras.* Santo Domingo, Dominican Republic: PRODECO.

Coyula, Mario. 1996. "The Neighborhood as Workshop." *Latin American Perspectives* 23 (4): 90–103.

Cruz, María Caridad. 1992. "Agricultura Urbana. Una Experiencia de aprovechamiento de los espacios disponibles en la ciudad de la Habana." Unpublished manuscript.

———. 1994. "La agricultura en la Habana: Evaluación de una experiencia." Paper prepared for *Georural* 94. Havana.

Cruz Pérez, Rafael. 1988. "Contaminación Producida por Explosivos y Residuous de Explosivos en Vieques, Puerto Rico." *Dimensión* 8, no. 2: 37–42.

Dávila, Arlene. 1997. *Sponsored Identities: Cultural Politics in Puerto Rico.* Philadelphia: Temple University Press.

Deere, Carmen Diana. 1993. "Cuba's National Food Program and Its Prospects for Food Security." *Agriculture and Human Values* 10: 35–51.

———. 1995. "The New Agrarian Reforms." *NACLA Report on the Americas* 24, no. 2: 13–17.

Deere, Carmen Diana, Mieke Meurs, and Niurka Pérez. 1992. "Toward a Periodization of the Cuban Collectivization Process: Changing Incentives and Response." *Cuban Studies* 22: 115–149.

Delgado, James. 1996. *Ghost Fleet: The Sunken Ships of the Bikini Atoll.* Honolulu: University of Hawaii Press.

Delgado Cintrón, Carmelo. 1989. *Culebra y la Marina de Estados Unidos.* Río Piedras, Puerto Rico: Editorial Edil.

De Lisio, Antonio. 1992. "La sustentabilidad: Nuevo ambientalismo o viejo desarrollismo?" In Andrés Serbin, ed., *Medio ambiente, seguridad, y cooperación regional en el Caribe.* Caracas: Coedition INVESP/CIQRO/Editorial Nueva Sociedad.

del Rosario, Pedro Juan. 1987. "Economia rural en la República Dominicana: Una nueva visión de los problemas agrarios." Paper presented at Seminario sobre Teledetección, Sistemas Agrarios y Degradación Ambiental, Universidad Católica Madre y Maestra, Santiago, R.D.

del Rosario, Pedro Juan, Helmut Schorgmayer, Frans Geilfus, Luc St. Pierre, and José Miguel Hernández. 1996. *Uso de la Tierra y Producción de Alimentos en la República Dominicana.* Santiago de los Caballeros, R.D.: Pontífica Universidad Catolica Madre y Maestra, Colección Estudios no. 180.

Departamento de Salud. 2000. *Resumen Demográfico.* Available at www.salud.gov.pr/estadisticas/Estadisticas/Resumen%20Demográfico%202000.htm.

Departamento de Salud. 2002. *Resumen Demográfico.* Available at www.salud.gov.pr.

Department of Natural and Environmental Resources. 1999. *Puerto Rico and the Sea*. San Juan, Puerto Rico.

Diamond, Jared. 2005. *Collapse: How Societies Choose to Fail or Succeed*. New York: Viking.

Díaz, Beatriz. 1995. "Biotecnología Agrícola: Estudio de Caso en Cuba." Paper prepared for the 19th International Congress of the Latin American Studies Association. Washington, D.C.

Díaz, Beatriz, and Marta Muñoz. 1994. "Biotecología agrícola y medio ambiente en el Periodo Especial Cubano." Paper prepared for the 18th Annual Congress of the Latin American Studies Association. Atlanta, Ga.

Díaz Briquets, Sergio, and Jorge Pérez-López. 2000. *Conquering Nature: The Environmental Legacy of Socialism in Cuba*. Pittsburgh: University of Pittsburgh Press.

Díaz Román, Miguel. 1999. "En entredicho origen del auge de la construcción." *El Nuevo Día*, September 12, Negocios, 10.

Dibblin, Jane. 1988. *Day of Two Suns: U.S. Nuclear Testing and the Pacific Islanders*. New York: New Amsterdam.

Dirección Nacional de Parques, Agencia Espanola de Cooperación Internacional, and the Agencia de Medio Ambiente, Junta de Andalucia. 1991. *Plan de uso y gestión del Parque Nacional de Los Haitises y Areas Periféricas. Documento Sintesis*. Santo Domingo: Editora Corripio.

Dirección Nacional de Parques and the Instituto de Cooperación Iberoamericana (ICI). 1989. "Plan de uso y gestión del Parque Nacional 'Los Haitises' y sus Areas Periféricas." *Parques Nacionales* 3: 4–7.

Dore y Cabral, Carlos. 1982. *Problemas de la Estructura Agraria Dominicana*. Santo Domingo: Ediciones Taller.

Douence, J. C. 1990. "Le droit de l'environnement dans les departements d'outre mer." Rapport general. Societe Francaise pour le Droit de l'Environnement.

Dryzek, John S. 1997. *The Politics of the Earth*. New York: Oxford University Press.

Duany, Jorge. 2000. "Nation on the Move: The Construction of Cultural Identity in Puerto Rico." *American Ethnologist* 27, no. 1 (February): 5–30.

Dubois, Alfonso. 1993. "Las organizaciones no gubernamentales en el debate sobre el desarrollo." *Papeles para Paz* [Madrid] 47/48. Centro de Investigación para la Paz (CIP).

Dunlap, Riley E., and Angela G. Mertig. 1992. "The Evolution of the U.S. Environmental Movement from 1970 to 1990: An Overview." In *American Environmentalism: The U.S. Environmental Movement, 1970–1990*, ed. Riley E. Dunlap and Angela G. Mertig. Washington, D.C.: Taylor and Francis.

Eckstein, Susan. 1994. *Back from the Future: Cuba under Castro*. Princeton, N.J.: Princeton University Press.

Economist, Intelligence Unit. 1994. *Country Report: Cuba, Dominican Republic, Haiti, Puerto Rico*. Fourth Quarter.

Economist. 1996. "Survey Cuba." April 6, 3–6.

El Expreso. 2004. "Identifican causas pobreza para Puerto Rico." October 7–13.

Enriquez, Laura J. 1994. *The Question of Food Security in Cuban Socialism. Exploratory Essays, no. 1*. Berkeley: International and Area Studies, University of California at Berkeley.

———. 2003. "Economic Reform and Repeasantization in Post-1990 Cuba." *Latin American Research Review* 38, no. 1: 202–218.

Environmental Quality Board. 1994. *Water Quality in Puerto Rico*. San Juan, Puerto Rico: Environmental Quality Board.

Estades-Font, María Eugenia. 1988. *La Presencia Militar de Estados Unidos en Puerto Rico, 1898–1918: Intereses Estratégicos y Dominación*. Río Piedras: Ediciones Huracán.

Esty, Donald, and Marion Chertow, eds. 1997. *Thinking Ecologically: The Next Generation of Environmental Policy*. New Haven, Conn.: Yale University Press.

Evanson, Debra. 1994. *Revolution in the Balance: Law and Society in Contemporary Cuba*. Boulder, Colo.: Westview Press.

FAO (Food and Agriculture Organization of the United Nations). 1991. *Forestry Action Plan for Dominican Republic*. Santo Domingo: Secretariado Tecnico de la Presidencia, Comisión Nacional Técnica Forestal.

Falconbridge. 2005. Our business: Nickel-Falcondo. www.falconbridge.com/our-business/nickel_falcondo.html.

Figueras, Miguel Alejandro. 1992. "The Transformation of the Cuban Sugar Complex." In The Cuban Revolution into the 1990s, ed. Centro de Estudios sobre America. Boulder, Colo.: Westview Press.

Figueroa, M. 2003. "Infantil el rostro de la pobreza." *El Nuevo Día*, August 27.

Fireside, Daniel. 2002. *Coffee Crisis and Opportunity: Making Guatemala's Coffee Economy Work for Small Farmers*. Master's thesis in Regional Planning, Cornell University.

Fisher, Julie. 1998. *Non-Governments: NGOs and the Political Development of the Third World*. West Hartford, Conn.: Kumarian Press.

Fiske, Shirley. 1992. "Sociocultural Aspects of Establishing Marine Protected Areas." *Ocean and Coastal Management* 17, no. 1:25–46.

Fitte-Duval, Annie, and Fred Reno. 1993. "Les associations et la defense de l'environnement: Le cas de l'ASSAUPAMAR a Ia Martinique." Colloque Espaces Naturels et Espaces Sociaux, BESSOR, Université de Rennes.

Fox, Jonathan, and Luis Hernandez. 1992. "Mexico's Difficult Democracy: Grassroots Movements, NGOs, and Local Government." *Alternatives* 17: 165–208.

Franks, Julie. 1997. *Transforming Property: Landholding and Political Rights in the Dominican Sugar Region, 1880–1930*. Unpublished Ph.D. dissertation, SUNY-Stonybrook.

Freudenberg, Nicholas. 1984. *Not In Our Backyards! Community Action for Health and the Environment*. New York: Monthly Press.

Friedmann, Harriet. 1995. "Food Politics: New Dangers, New Possibilities." In *Food and Agrarian Orders in the World-Economy*, ed. Philip McMichael. New York: Praeger.

Gandy, Matthew. 2002. *Concrete and Clay: Reworking Nature in New York City*. Cambridge, Mass.: MIT Press.

GAPT (Grupo de Apoyo Técnico y Profesional para el Desarrollo Sustentable de Vieques). 2000. "Guidelines for the Sustainable Development of Vieques."

———. 1994. "Efectividad simbolica, practicas sociales, estrategias del movimiento ambientalista Venezolano: Sus impactos en la democracia." In *Retos para el desarrollo y la democracia movimientos ambientales en America Latina y Europa*, ed. Maria-Pilar García Gaudilla and Jutta Blauert. Caracas: Fundación Friedrich Ebert de Mexico/Editorial Nueva Sociedad.

García, Ilia. 1993. "Movimientos sociales regionales y construcción de identidades regionales." Paper presented at the International Symposium, "Diversidad cultural y construcción de identidades en America Latina y el Caribe: nuevos enfoques," Caracas, Venezuela.

García, Neftalí. 1993. "Ponencia sobre Alternativas Energéticas para PR, Asamblea Colegio de Abogados en Mayagüez." Unpublished document.

———. 1998. "Tecnologías Alternas para Lidiar con el problema de los Desperdicios Sólidos." Unpublished document.

———. 2002. "Factores Limitantes en el Desarrollo de Sistemas Integrados de Manejo de Residuos Sólidos en Puerto Rico." Unpublished document.

———. 1988. "Sobre la organización comunal." *Pensamiento Crítico* 16, no. 69: 2–15.

García Canclini, Nestor. 1999. *La globalización imaginada.* Buenos Aires: Editorial Paidós.

García-Guadilla, María-Pilar. 1991. "La restructuración del movimiento ambientalista en Venezuela: Tipología y perspectivas políticas." In *Ambiente, estado y sociedad,* ed. Maria-Pilar García Guadilla. Caracas: Universidad Simon Bolivar, Centro de Estudios del Desarrollo (CENDES-UCV).

García-Guadilla, María-Pilar, and Jutta Blauert, eds. 1994. *Retos para el desarollo y la democracia: movimientos ambientales en América Latina y Europa.* Caracas, Venezuela: Editorial Nueva Sociedad.

García-Martinez, N. 1972. *Puerto Rico y la Minería.* San Juan: Librería Internacional.

———. 1975. "Puerto Rico's Colonial Economy in the XX Century: Preliminary Notes." Unpublished document.

———. 1976a. "La contaminación ambiental en Puerto Rico." Unpublished document.

———. 1978. "Puerto Rico Siglo XX: Lo histórico y lo natural en la ideología colonialista." *Pensamiento Crítico* 8.

———. 1983a. "Puerto Rico: Presencia y efectos de las compañías transnacionales." Unpublished document.

———. 1983b. "Puerto Rico: Crisis and Alternatives." Unpublished document.

———. 1984a. "Apuntes para una historia de la lucha ambiental." Unpublished document.

———. 1984b. "Economía política de los problemas ambientales." Unpublished document.

García-Muñiz, Humberto. 2001. "Goliath against David: The Battle for Vieques as the Last Crossroad?" *Centro Journal* (spring): 126–141.

Geilfus, Franz. 1985. *Estudio de Impacto Ambiental de Alternativas Agroforestales en Los Haitises. Informe para la realización del Plan de Manejo Parque Nacional Los Haitises.* Santo Domingo: Dirección Nacional de Parques.

Ghai, Daram, Cristóbal Kay, and Peter Peek. 1988. *Labor and Development in Rural Cuba.* London: Macmillan.

Gilbe, Carlos J. 1998. "El manejo de las cuencas hidrológicas en Puerto Rico: La autonomía municipal frente a la crisis del agua potable en el Area Metropolitana de San Juan." *Ambiente y Desarrollo en el Caribe (ADEC)* Boletín Electrónico (http://adec.upr.clu.edu), Centro de Investigaciones Sociales, Universidad de Puerto Rico, Recinto de Río Piedras.

Giusti, Juan A. 1994. *Labor, Ecology and History in a Caribbean Sugar Plantation Region: Piñones (Loiza), Puerto Rico, 1770–1950.* Unpublished Ph.D. dissertation, State University of New York at Binghamton.

———. 1999. "La Marina en la mirilla: Una comparación de Vieques con los campos de bombardeo y adiestramiento en los Estados Unidos." In *Fronteras en conflicto:*

Guerra contra las drogas, militarización y democracia en el caribe, Puerto Rico y Vieques, ed. Humberto García Muñiz and Jorge Rodríguez Beruff. San Juan: Red Caribeña de Geopolítica, Cuadernos de Paz 1.

———. 2001. "Política ambiental: ¿Síntesis y práctica de las ciencias sociales?" In *Ciencias sociales: Sociedad y cultura contemporáneas*, 2d edition, ed. Lina M. Torres Rivera. México, D.F.: International Thomson.

Goffman, Erving. 1986. *Frame Analysis*. Boston: Northeastern University Press.

Goldman, Benjamin A., and Laura Fitton. 1994. *Toxic Wastes and Race Revisited: An Update of the 1987 Report on the Racial and Socio-Economic Characteristics of Communities with Hazardous Waste Sites*. New York: Center for Policy Alternatives, National Association for the Advancement of Colored People, and United Church of Christ Commission for Racial Justice.

Goldman, Michael. 2001a. "The Birth of a Discipline: Producing Authoritative Green Knowledge, World Bank Style." *Ethnography* 2: 191–217.

———. 2001b. "Imperial Science, Imperial Nature: Environmental Knowledge for the World (Bank)." In *Earthly Politics*, ed. Sheila Jasanoff and Marybeth Long Martello. Ithaca, N.Y.: Cornell University Press.

Gonzalez, David. 2001. "Vieques Voters Want the Navy to Leave Now." *New York Times*, September 16.

———. 2002. "Aguacate Journal: Cuba's Bittersweet Move to Trim Its Sugar Crop." *New York Times*, October 9, Section A, p. 4, col. 3.

González, Raymundo. 1993. "Ideologia del progreso y campesinado en el siglo XIX." *Ecos* 1 (2): 25–76.

González Martínez, Alfonso. 1994. "Las luchas ecológico-sociales en Mexico: Prospectivas." In *Retos para el desarrollo y la democrácia: Movimientos ambientales en America Latina y Europa*, ed. María-Pilar García Gaudilla and Jutta Blauert. Caracas: Fundación Friedrich Ebert de Mexico/Editorial Nueva Sociedad.

Gottdiener, Mark. 1985. *The Social Production of Space*. Austin: The University of Texas Press.

Gouldner, Alvin W. 1985. *El Futuro de los Intelectuales y las Nuevas Clases en Ascenso*. México, D.F.: Alianza Editorial.

Griffith, David C. 1999. *The Estuary's Gift: An Atlantic Coast Cultural Biography*. University Park: The Pennsylvania State University Press.

Griffith, David C., E. Kissam, J. Camposeco, A. García, M. Pfeffer, D. Runsten, and M. Valdés Pizzini. 1995. *Working Poor: Farmworkers in the United States*. Philadelphia: Temple University Press.

Grosfoguel, Ramón. 2003. *Colonial Subjects: Puerto Ricans in a Global Perspective*. Berkeley: University of California Press.

Gudynas, Eduardo. 1992. "Los multiples verdes del ambientalismo latinoamericano." *Nueva Sociedad* [Caracas] 122.

Gudynas, Eduardo, and Graciela Evia. 1991. *La praxis por la vida: Introduccióna metodologías de la ecología social*. Montevideo: CIPFE/CLAES/NORDAN.

Guimaraes, Roberto. 1992. "El discreto encanto de la Cumbre de la Terra. Evaluación impresionista de Rio–92." *Nueva Sociedad* [Caracas] 122.

Guadalupe-Fajardo, Evelyn. 2000. "Forecast: This Winter Season Will Be Hot." *Caribbean Business*, Thursday June 15, 20–26.

Guptill, Amy. 1996. "The Changing Structure of Cuban Agriculture in the Post-Soviet World." Unpublished manuscript, Cornell University.

Gutiérrez, Elías. 2000. *El futuro sobre el tapete: Proyecto Delphi*. San Juan, Puerto Rico: Cuadernos del Centro de Investigación y Política Pública (CIPP).

Gutiérrez, E. R. 1996. "Necesidades Prioritarias del Puerto Rico que Emerge." http://egp.rrp.upr.edu/Investigación/ERGPRIO.htm.

Guzmán-Ríos, S., and Quiñones-Márquez, F. 1985. "Reconnaissance of Trace Organic Compounds in Ground Water throughout Puerto Rico, October 1983." U.S. Geological Survey Open-File Data Report 84–810, 1 sheet.

Hall, Covell. 1995. "Saving the Fish." *Jamaica Weekly Gleaner*, November 17–23.

Hall, Stuart. 1991. "Old and New Ethnicities." In Anthony King, ed., *Culture, Globalisation and the World System: Contemporary Conditions for the Representation of Identity*. London: MacMillan.

Harris, Lis. 2003. *Tilting at Mills: Green Dreams, Dirty Dealings, and the Corporate Squeeze*. Boston: Houghton-Mifflin.

Hartlyn, Jonathan. 1998. *The Struggle for Democratic Politics in the Dominican Republic*. Chapel Hill: University of North Carolina Press.

Harvey, David. 1996. *Justice, Nature, and the Politics of Difference*. Malden, MA: Blackwell.

———. 1999. "The Environment of Justice." In *Living with Nature: Environmental Politics as Cultural Discourse*, ed. Frank Fischer and Maarten Hajer, 153–185. New York: Oxford University Press.

Henry-Wilson, Maxine A. 1990. "Community Empowerment: Alternative Structures for National Development." In *Integration and Participatory Development*, ed. Judith Wedderburn. Jamaica: Friedrich Ebert Stiftung.

Hernández, Marialba. 1990. "Las organizaciones ambientalistas comunitarias en la costa oeste de Puerto Rico: Su origen y desarrollo." Sociology Seminar Paper on Coastal Gentrification. Department of Social Sciences, University of Puerto Rico, Mayagüez.

Hernandez, Raymond. 2001. "Vieques Issue Is Put on Hold in Response to Terrorism." *New York Times,* September 27, 2001.

Hershkowitz, Allen. 2003. *Bronx Ecology: Blueprint for a New Environmentalism*. Washington, D.C.: Island Press.

Hexner, Thomas J., and Glenn Jenkins. 1998. "Puerto Rico: The Economic and Fiscal Dimensions." Report prepared for Citizens Educational Foundation.

Hiller, Herbert L. 1979. "Tourism: Development or Dependence?" In *The Restless Caribbean*, ed. Richard Millett and Will W. Marvin. New York: Praeger.

Hoetink, Harry. 1982 [1972]. *The Dominican People, 1850–1900: Notes for a Historical Sociology*. Trans. S. K. Ault. Baltimore: Johns Hopkins University Press.

Hojnian, David K. 1993. "Non-Governmental Organizations (NGOs) and the Chilean Transition to Democracy." *European Review of Latin America and Caribbean Studies* 54 (June).

Holder, Jean S. 1993. "The Caribbean Tourism Organization in Historical Perspective." In Dennis J. Gayle and Jonathan N. Goodrich, *Tourism Marketing and Management in the Caribbean*. London: Routledge.

Honey, Martha. 1999. *Ecotourism and Sustainable Development: Who Owns Paradise?* Washington, D.C.: Island Press.

Hunter, John M., and Sonia I. Arbona. 1995. "Paradise Lost: An Introduction to the Geography of Water Pollution in Puerto Rico." *Social Science and Medicine* 40, no. 10: 1331–1355.

Hurrell, Andrew. 1992. "El medio ambiente y las relaciones intenacionales, una perspectiva mundial." In *Medio anbiente y relaciones internacionales*, ed. Guhl, Emesto, and Juan G. Tokatlian. Santafe de Bogóta: Tercer Mundo.

Iglesias, Enrique. 1993. "El papel de los organismos multilaterales de cooperacíon en el desarrollo sostenible: El caso del BID." *Sintesis* [Madrid] 20 (July-December).

Institute of Medicine, Environmental Justice, Health Sciences Policy Program, Health Sciences Section. 1999. *Toward Environmental Justice: Research, Education and Health Policy Needs*. Washington, D.C.: National Academy Press.

Inter-American Development Bank. 1994. *A Focus on Participation: Fifth Consultative Meeting on the Environment*. Washington, D.C.: Inter-American Development Bank.

Iranzo, Guillermo. 1996. *De la práctica de la pesca a la práctica del ocio: Desarrollo turístico y privatización de tierras litorales en la Antilla Menor de San Ildefonso de Culebra. Uso de la tierra en Culebra.* Unpublished Ph.D. dissertation, Universidad Autónoma de México.

Iranzo Berrocal, Guillermo. 1994. "Aproximación Antropológica al Turismo y la Gentrificación de la Tierra en Culebra, Puerto Rico." Paper delivered at a conference in Cancún, México. Archivo Histórico, Fuerte Conde de Mirasol, Vieques, Puerto Rico.

Irizarry-Mora, Edwin. 1996. "Programa del PIP para el ambiente." *El Nuevo Día* [San Juan, Puerto Rico] 20: 71.

Island Resources Foundation. 1991. Directory of Environmental NGOs in the Eastern Caribbean. St. Thomas, U.S. Virgin Islands, April.

Jacobeit, Cord. 1996. "Nonstate Actors Leading the Way: Debt-for-Nature Swaps." In *Institutions for Environmental Aid*, ed. Robert O. Keohane and Mark A. Levy. Cambridge. Mass.: MIT Press.

Jácome, Francine. 1995. "Los movimientos ambientales y la cooperación ambiental en el Caribe: Una primera aproximación." Paper presented at the International Seminar "Hacia una Agenda Socio-política de la Integración en el Caribe," organized by FLACSO-Dominican Republic, Santo Domingo, Dominican Republic, March.

Jácome, Francine, coord. 1996. *Retos de la cooperación ambiental: el caso del Caribe*. Caracas, Venezuela: Editorial Nueva Sociedad.

Jácome, Francine, and Glenn Sankatsing. 1992. "La cooperación ambiental en el Caribe: Actores principales." In *Medio ambiente, seguridad, y cooperación regional en el Caribe*, ed. Andrés Serbin. Caracas: Coedición INVESP/CIQRO/Editorial Nueva Sociedad.

Jamaica Weekly Gleaner. 1992. "MoBay's Marine Park to Be Officially Opened." July 10, p. 8.

James, Sherman A., Amy J. Schulz, and Juliana van Olphen. 2001. "Social Capital, Poverty, and Community Health: An Exploration of Linkages." In *Social Capital and Poor Communities*, ed. Susan Saegert, J. Phillip Thompson, and Mark P. Warren, 165–188. Ford Foundation Series on Asset Building. New York: Russell Sage Foundation.

Jean-Baptiste, Neudy. 1999. *Distribución espacial de las nasas y sus relaciones con la topografía, las agregaciones de peces y las capturas estacionales.* Master's thesis, University of Puerto Rico at Mayagüez, Department of Marine Sciences.

Katinas, Paula. 1998. "Is Gowanus Causing Kids' Asthma?" *Home Reporter and Sunset News* 46, no. 2 (January 23): 2ff.

Kearney, A. T., Management Consultants. 2004. *Puerto Rico 2025 Project Assessment*,

Report on Current Status of the Economic, Social, Environmental and Infrastructure Development in Puerto Rico. Chicago: A. T. Kearney, Inc.

Keck, Margaret. 1995. "Social Equity and Environmental Politics in Brazil: Lessons from the Rubber Tappers of Acre." *Comparative Politics* (July): 409–21.

Keck, Margaret, and Kathryn Sikkink. 1998. *Activists Beyond Borders: Advocacy Networks in International Politics.* Ithaca, N.Y.: Cornell University Press.

Keita, Yasmina. 1903. *Le littoral domient: Droit et politique.* M.A. thesis, CREJETA, Université des Antilles et de la Guyane.

Kincaid, Jamaica. 1988. *A Small Place.* New York: Farrar, Straus, Giroux.

Kiste, Robert. 1974. *The Bikinians: A Study in Forced Migration.* Menlo Park, Calif.: Cummings Publishing Company.

Klein, Deborah. 2001. "For the Future of Vieques, Look to Hawaii." *New York Times,* June 16.

Knox, Paul. 1993. "Capital, Material Culture and Socio-spatial Differentiation." In *The Restless Urban Landscape,* ed. Paul L. Knox. Englewood Cliffs, N.J.: Prentice Hall.

Koont, Sinan. 2004. "Food Security in Cuba." *Monthly Review* 55 (8): 11–30.

Krausse, Gerald H. 1994. "La Parguera, Puerto Rico: Balancing Tourism Conservation and Resource Management." Paper presented at the World Congress on Tourism for the Environment, Puerto Rico, May 31–June 4, 1994.

Langley, Lester. 1985. "Roosevelt Roads, Puerto Rico, U.S. Naval Base 1941–." In Paolo Coletta with K. Jack Bauer, *United States Navy and Marine Corps Bases, Overseas.* Westport, Conn.: Greenwood Press.

Lara Fontánez, Juan. 1998. "Some Economic Issues for Puerto Rican Leaders in the (Early) 21st Century." *Conferencias y Foros* Número 46, Unidad de Investigaciones Económicas, Universidad de Puerto Rico en Río Piedras.

Ledogar, Robert J., Analia Penchaszadeh, C. Cecilia Iglesias Garden, and Luis Garden Acosta. 2000. "Asthma and Latino Cultures: Different Prevalence Reported among Groups Sharing the Same Environment." *American Journal of Public Health* 90: 929–935.

Lee, Moon Wha. 1999. *Housing New York City 1996.* New York.

Leff, Enrique. 1992. "Environment and Democracy." In *Democratic Culture and Governance, UNESCO/Ediciones Hispamerica,* coordinated by Luis Albala-Bertrand. College Park: Latin American Studies Center, University of Maryland.

———. 1993. "Analisis sociológico del movimiento ambientalista en América Latina." In *Ambiente, estado y sociedad,* ed. María-Pilar García Guadilla. Caracas: Universidad Simon Bolivar/Centro de Esuclios del Desarrollo [CENDES].

———. 1994. "El movimiento ambiental y las perspectivas de la democracia en America Latina." In *Retos para el desarrollo y la democrácia: Movimientos ambientales en América Latina y Europa,* ed. María Pilar García Guadilla and Jutta Blauert. Caracas: Fundación Friedrich Ebert de Mexico/Editorial Nueva Sociedad.

Leis, Hector. 1992. "El rol educativo del ambientalismo en la política mundial." *Nueva Sociedad* (Caracas) 122 (November-December).

Levins, Richard. 1993. "The Ecological Transformation of Cuba." *Agriculture and Human Values* 10, no. 3: 52–60.

Levins, Richard, and Richard Lewontin. 1985. *The Dialectical Biologist.* Cambridge: Harvard University Press.

Levy, Horace. 1987. "Community-Based Ecotourism in Jamaica." *Caribbean Park and Protected Area Bulletin* 5, no. 1: 8.

Lewis, David K. 1990. "Non-Governmental Organizations and Alternative Strategies: Bridging the Development Gap between Central America and the Caribbean." In *Integration and Participatory Development*, ed. Judith Wedderburn. Kingston, Jamaica: Friedrich Ebert Foundation.

———. 1994. "La propuesta para el establecimiento de la Asociación de Estados del Caribe (AEC): una evaluación prospective." Lecture presented at the seminar/workshop "La Asociación de Estados del Caribe," Caracas.

Lezama, Cecilia. 2000. *Contaminación ambiental y estrategias empresariales de cambio tecnológico en la industria galvanizadora de Guadalajara*. Masters thesis, CIESAS/ Guadalajara.

López Montañez, Wilfredo, and Marianne Meyn. 1992. "Desarrollo colonial y destrucción ambiental." *Claridad, En Rojo*, January 17–23, pp. 12–13.

Lynch, Barbara. 1992. "The Yautia Boom and the Sustainability of Conuco Cultivation in the Dominican Republic." Paper prepared for the 4th North American Symposium on Society and Natural Resource Management, Madison, Wisconsin, May.

———. 2001. "Development and Risk: Urban Environmental Discourse and Danger in Dominican and Cuban Urban Watersheds." In *International Political Economy of the Environment, International Political Economy Yearbook*, Volume 12, ed. Dimitris Stevis and Valerie Assetto. Chapter 8. Boulder, Colo.: Lynne Reiner.

Macdonald, Laura. 1992. "Turning to the NGOs: Competing Conceptions of Civil Society in Latin America." Paper presented at the Latin American Studies Association (LASA) Annual Meeting, Los Angeles, September.

———. 1997. *Supporting Civil Society: The Political Role of Non-governmental Relations in Central America*. New York: St. Martin's Press.

Maldonado, Marta M. 2000. "No al Carbón! Reflections on Technology, Environment, and Community Empowerment in Puerto Rico." In *Chicana/o Latina/o Studies for the 21st Century: New Perspectives on Mentorship and Research*, ed. Marcos Pizarro. Pullman, Wash.: Writings from Encuentros in C/L Studies: A Speaker's Series at Washington State University.

Mariñez, Pablo. 1984. *Resistencia Campesina, Imperialismo y Reforma Agraria en República Dominicana (1899–1978)*. Santo Domingo: Ediciones CEPAE.

Marmora, Leopoldo. 1992. "Del Sur explotado al Sur marginado: Justicia económica y justicia ecológica a escala global." *Nueva Sociedad* (Caracas) 122.

Márquez, Lirio, and Jorge Fernández. 2000. "Environmental and Ecological Damage to the Island of Vieques Due to the Presence and Activities of the United States Navy." Special International Tribunal on the Situation of Puerto Rico and the Island Municipality of Vieques.

Martínez, Eleuterio. 1993. *Cumbre para la Tierra: Perspectivas Mundiales*. Santo Domingo: Foro Urbano.

Martínez, Ramón. 1994. "Status del manejo y reglamentación de los manglares en Puerto Rico." In *El Ecosistema de Manglar en América Latina y la Cuenca del Caribe: Su Manejo y Conservación*, ed. Daniel O. Suman. New York: The Tinker Foundation.

McCaffrey, Katherine T. 2002. *Military Power and Popular Protest: The U.S. Navy in Vieques, Puerto Rico*. New Brunswick, N.J.: Rutgers University Press.

McElroy, Jerome, and Klaus de Albuquerque. 1991. "Tourism Styles and Policy Responses in the Open Economy-Closed Environment Context." In *Caribbean Ecology and Economics*, ed. Norman P. Girvan and David A. Simmons. St. Michaels, Barbados: Caribbean Conservation Association.

McGregor, Duncan F.M., David Barker, and Sally Lloyd Evans, eds. 1998. *Resource Sustainability and Caribbean Development*. Mona: The Press University of the West Indies.

McIntyre, Lionel, and Rudy Dupuy. 1996. "Vieques Island, Puerto Rico: A Development Strategy for the Naval Ammunitions Facility." Vieques Studio Urban Planning Program, Columbia University.

McLaren, Deborah. 1998. *Rethinking Tourism and Ecotravel: The Paving of Paradise and What You Can Do to Stop It*. West Hartford, Conn.: Kumarian Press.

Mendoza, Fernando S., and Elena Fuentes-Afflick. 1999. "Latino Children's Health and the Family-Community Health Promotion Model." *The Western Journal of Medicine* 170: 85–92.

Millán-Pabón, C. 2003. "Traba económica: la pobreza infantil." *El Nuevo Día*, August 27, sec. El País.

Miller, Marian A. L. 1992. "Balancing Development and Environment: The Third World in Global Environmental Politics." *Society and Natural Resources* 5: 297–305.

———. 2001. "Tragedy for the Commons: The Enclosure and Commodification of Knowledge." In Dimitris Stevis and Valerie Assetto, eds., *International Political Economy of the Environment*, International Political Economy Yearbook, Volume 12. Chapter 6. Boulder, CO: Lynne Reiner

Ministry of Science, Technology and Environment (CITMA). 1996. *Programa Nacional de Medio Ambiente y Desarrollo*. La Havana: Centro de Información, Divulgación y Educación Ambiental (CIDEA), Agencia de Medio Ambiente. (Published in co-operation with Worldwide Fund for Nature.)

Mintz, Sidney. 1985. *Sweetness and Power: The Place of Sugar in Modern History*. New York: Viking

Montes, Marisol, and Wanda I. Santana. 1994. "Conflicto político-ecológico en Guánica: Rezonificación del sector Cerro San Jacinto en el Bosque Seco y su reserva natural." Political Science Seminar Papers, Department of Social Sciences, University of Puerto Rico, Mayagüez.

Morris, Nancy. 1995. *Puerto Rico: Culture, Politics, Identity*. Westport, Conn.: Praeger Publishers.

Mowforth, Martin, and Ian Mount. 1998. *Tourism and Sustainability: New Tourism in the Third World*. London: Routledge.

Munasinghe, Viranjini. 2001. *Callaloo or Tossed Salad? East Indians and the Cultural Politics of Identity in Trinidad*. Ithaca, N.Y.: Cornell University Press.

Muñoz, Myra. 2001. "Vieques Movement Has Attracted People from All Walks of Life." *San Juan Star*, July 1.

Murray, Douglas L. 1994. *Cultivating Crisis: The Human Cost of Pesticides in Latin America*. Austin: University of Texas Press.

Murray, John A. 1991. *The Islands and the Sea: Five Centuries of Nature Writing from the Caribbean*. New York: Oxford University Press.

Navajas, Hugo, et al. 1997. *Environmental Programmes in Latin America and the Caribbean: An Assessment of UNDP Experience*. New York: United Nations Development Program.

Navarro, Ana, and Eduardo Navarro. 2001. *Guía ambiental para Puerto Rico*. University of Puerto Rico Sea Grant College Program, University of Puerto Rico at Mayagüez.

National Research Council. 1989. *Alternative Agriculture*. Washington, D.C.: National Academy Press.

Neggers, Xaviera. 1995. "Construction Boomed in '95." *The San Juan Star,* San Juan, Puerto Rico, October 28, p. B29.

Nettleford, Rex M. 1985. *Dance Jamaica: Cultural Definition and Artistic Discovery: The National Dance Theatre Company of Jamaica, 1962–1983.* New York: Grove Press.

Neufville, Zade. 1993. "Blue Mt. Park Changing Lives." *Jamaica Weekly Gleaner*, February 5, p. 16.

New York City Department of Health. 1999. *Asthma Facts.* New York: Community HealthWorks.

———. 2003. *Neighborhood Health Profiles, 2000.* Accessed August 2003.

New York State Department of Environmental Conservation. 1984. *Small Quantity Generators in New York State.* December.

New York Times. 2000. "Chipping Away at Embargo Reflects a New Vision of Cuba." *New York Times,* July 22, p. 3.

Nieves Falcón, Luis, Pablo García Rodríguez, and Félix Ojeda Reyes. 1971. *Puerto Rico Grito y Mordaza.* Río Piedras, Puerto Rico: Ediciones Librería Internacional.

Novotny, Patrick. 2000. *Where We Live, Work and Play: The Environmental Justice Movement and the Struggle for a New Environmentalism.* Praeger Series in Transformational Politics and Political Science. Westport, Conn.: Praeger.

Paniagua Pascual, María. 1998. *Environmental Discourses and Collective Action in the Dominican Republic.* Master's thesis, City and Regional Planning, Cornell University.

Patullo, Polly. 1996. *Last Resorts: The Cost of Tourism in the Caribbean. London:* Cassel Wellington House.

Pearce, Neil. 1996. "Traditional Epidemiology, Modern Epidemiology, and Public Health." *American Journal of Public Health* 86: 678–683.

Peet, Richard, and Michael Watts, eds. 2004. *Liberation Ecologies: Environment, Development, Social Movements*, 2d ed. New York and London: Routledge.

Pellow, David Naguib. 2002. *Garbage Wars: The Struggle for Environmental Justice in Chicago.* Cambridge: MIT Press.

Pérez, César. 1996. *Urbanización y Municipio en Santo Domingo.* Santo Domingo: Instituto Tecnologico de Santo Domingo.

Pérez, Ricardo. 2000. *Fragments of Memory: The State and Small-Scale Fisheries Modernization in Southern Puerto Rico.* Ph.D. dissertation, Anthropology, University of Connecticut at Storrs.

Perez-Pena, Richard. 1994. "State Admits Plants Headed to Poor Areas." *New York Times*, March 15.

Peters, Phillip. 2000. "Using Market Mechanisms to Deliver Affordable Food." *Cuba Today* 2 (2): 5–6.

Ploeg, Jan Douwe van der. 1990. *Labor, Markets, and Agricultural Production.* Trans. Ann Long. Boulder, Colo.: Westview Press.

Portes, Alejandro. 2000. "The Two Meanings of Social Capital." *Sociological Forum* 15, no. 1: 1–12.

Pumarada-O'Neill, Luís. 1993. *La Industria Cafetalera de Puerto Rico, 1736–1969.* San Juan: Oficina de Preservación Histórica.

Putnam, Robert D. 1995. "Bowling Alone: America's Declining Social Capital." *Journal of Democracy* 6 (January): 65–78.

———. 2000. *Bowling Alone: The Collapse and Revival of American Community.* New York: Simon and Schuster.

Ragster, LaVerne. 1993. "An Overview of the Consortium of Caribbean Universities for Natural Resource Management." St. Thomas: Working Paper of the Eastern Caribbean Center of the University of the U.S. Virgin Islands.

Ragster, LaVerne, and Lloyd Gardner. 1993. "Environmental Conservation Policies and Cooperation in the Caribbean." Paper presented at the Seminar "Medio ambiente en el caribe y el impacto de los factores socio-culturales," Caracas, November.

Ramírez, Rafael L. 1976. *El Arrabal y la Política*. Río Piedras: Editorial de la Universidad de Puerto Rico.

Ramirez, Roberto R., and G. Patricia de la Cruz. 2002. *The Hispanic Population in the U.S.: March 2002. Current Population Reports P20–545*. Washington, D.C.: U.S. Census Bureau.

Raynolds, Laura. 1994. "The Restructuring of Third World Agroexports: Changing Production Relations in the Dominican Republic." In *The Global Restructuring of Agro-Food Systems*, ed. Philip McMichael, 214–237. Ithaca, N.Y.: Cornell University Press.

———. 2000. "Re-embedding Global Agriculture: The International Organic and Fair Trade Movements." *Agriculture and Human Values* 17: 297–309.

Reilly, Charles A., ed. 1995. *New Paths to Democratic Development in Latin America: The Rise of NGO-Municipal Collaboration*. Boulder, Colo.: Lynne Rienner.

Renard, Yves, and Manuel Valdés Pizzini. 1994. "Hacia la participación comunitaria: Lecciones y principios guía en el Caribe." In *Actas de la Conferencia: La Participación Comunitaria en la Gestión Ambiental y el Co-manejo en la República Dominicana*. Santo Domingo, Dominican Republic: Instituto Caribeño de Recursos Naturales (CANARI) y el Centro para la Conservación y Ecodesarrollo de la Bahía de Samaná y su Entorno (CEBSE).

Renner, Michael. 1994. "Military Mop-Up." *WorldWatch* 7, no. 5 (September-October).

Rhodes, Eduardo Lao. 2003. *Environmental Justice in America: A New Paradigm*. Bloomington: Indiana University Press.

Rivera, Odalys. 1990. "El dilema ambiental en Puerto Rico: 20 años de lucha." *Dialogo*, April.

———. 1995. "Una nueva fuerza social en el ambiente." *Diálogo*, December.

Rivera Torres, Leticia, and Antonio J. Torres. 1996. "Vieques, Puerto Rico: Economic Conversion and Sustainable Development." Archivo Histórico, Fuerte Conde de Mirasol, Vieques, Puerto Rico.

Roberts, J. Timmons, and Nikki Demetria Thanos. 2003. *Trouble in Paradise: Globalization and Environmental Crisis in Latin America*. New York and London: Routledge.

Roberts, J. Timmons, and Melissa M. Toffolon-Weiss. 2001. *Chronicles from the Environmental Justice Frontline*. New York: Cambridge University Press.

Rocheleau, Dianne, and Laurie Ross. 1995. "Trees as Tools, Trees as Text: Struggles over Resources in Zambrana-Chacuey, Dominican Republic." *Antipode* 27: 407–428.

Rodríguez-Beruff, Jorge. 1988. *Política militar y dominación: Puerto Rico en el contexto Latinoamericano*. Río Piedras: Ediciones Huracán.

Rogers, Caroline S., Gilberto Cintrón, and Carlos Goenaga. 1978. "The Impact of Military Operations on the Coral Reefs of Vieques and Culebra." Report submitted to the Department of Natural Resources, San Juan, Puerto Rico.

Rosenbaum, Walter A. 2002. *Environmental Politics and Policy*. 5th ed. Washington, D.C.: Congressional Quarterly Press.

Rosenstreich, David L., Peyton Eggleston, Meyer Kattan, Dean Baker, Raymond G.

Slavin, Peter Gergen, Herman Mitchell, Kathleen McNiff-Mortimer, Henry Lynn, Dennis Ownby, and Floyd Malveaux for the National Cooperative Inner-City Asthma Study. 1997. "The Role of Cockroach Allergy and Exposure to Cockroach Allergen in Causing Morbidity among Inner-City Children with Asthma." *New England Journal of Medicine* 336, no. 19: 1356–1363.

Rosset, Peter, and Medea Benjamin. 1994. *Two Steps Back, One Step Forward: Cuba's National Policy for Alternative Agriculture.* Sustainable Agriculture Programme. Gatekeeper Series no. 46. London: International Institute for Environment and Development.

Rosset, Peter, with Shea Cunningham. 1994. "The Greening of Cuba." *Food First Action Alert* (spring), Institute for Food and Development Policy.

Safa, Helen I. 1974. *The Urban Poor of Puerto Rico: A Study in Development and Inequality.* Case Studies In Cultural Anthropology. New York: Holt, Reinhart and Winston.

Saltalamacchia, Homero R. 1995. La educación como política: Contextos y tareas del movimiento ambientalista en Puerto Rico. Hato Rey: Cuadernos del CENIAC, Universidad del Sagrado Corazón.

San Juan Star. 2001a. "Environmentalists Decry Land Transfer Terms." April 29.

———. 2001b. "Vieques Faces New Land Use Questions." May 6.

———. 2001c. "Vieques Movement Has Attracted People from All Walks of Life." July 1.

San Miguel, Pedro L. 1997. *Los campesinos del Cibao: Economía del mercado y transformación agraria en la República Dominicana, 1880–1960.* San Juan, P.R.: Editorial de la Universidad de Puerto Rico.

Scarano, F. A. 2000. *Puerto Rico Cinco Siglos de Historia.* 2d ed. New York: McGraw-Hill.

Schroeder, Richard A. 1993. "Shady Practice: Gender and the Political Ecology of Resource Stabilization in Gambian Garden/Orchards." *Economic Geography.*

Sears, Robin. Background Paper of the Task Force on Environmental Sustainability. UNDP, Millennium Project. www.unmillenniumproject.org/documents/tf06_july10.pdf.

Sepúlveda, Anibal, and Jorge Carbonell. 1988. *Cangrejos-Santurce: Historia ilustrada de su desarrollo urbano (1519–1950).* San Juan: Centro de Investigaciones CARIMAR, Oficina de Preservación Histórica.

Serbin, Andrés. 1992. "Seguridad ambiental y cooperación regional: Paradigmas, supuestos, percepciones, y obstáculos." In Medio ambiente, seguridad. y cooperacion regional en el Caribe, ed. Andres Serbin. Caracas: Coedición INVESP/CIQRO/Editorial Nueva Sociedad.

———. 1994. "Transnational Relations and Regionalization in the Caribbean." *Annals of the American Academy of Political and Social Science* 533.

———. 1996. El ocaso de las isles: el Gran Caribe frente a los desafíos globales y regionales. Caracas, Venezuela: Instituto Venezolano de Estudios Sociales y Políticos (INVESP) and Editorial Nueva Sociedad.

Sexton, Joe. 1994. "An Inquiry into Contamination Starts at the Brooklyn Navy Yard." *New York Times,* March 15.

Shanahan, John J. 2001. Declaration of John Shanahan. *The Commonwealth of Puerto Rico v. Hon. Donald Rumsfeld et al.* United States District Court for the District of Columbia.

Sharpe, Kenneth, and Barbara Lynch. 1993. "Environment and Development Programming in the Caribbean: Some Institutional Lessons from the Dominican Republic." Unpublished report to the Ford Foundation.

Sheller, Mimi. 2003. *Consuming the Caribbean: From Arawaks to Zombies.* New York: Routledge.

Shulman, Seth. 1992. *The Threat at Home: Confronting the Toxic Legacy of the U.S. Military.* Boston: Beacon Press.

Sierra Club. 2004. "Latino Communities at Risk: How Bush Administration Policies Harm Our Communities." www.sierraclub.org/comunidades.

Sinclair, Minor, and Martha Thompson. 2001. "Cuba Going against the Grain: Agricultural Crisis and Transformation." *Oxfam America Report.* Boston: Oxfam.

Smith, Allan. 1993. "Summary of the Activities of the Caribbean Natural Resources Institute in Relation to Integrated Rural Development in Coastal Areas." *Out of the Shell* (St. Lucia) 3, no. 1 (February).

Smith, Neil. 1996. *The New Urban Frontier: Gentrification and the Revanchist City.* London: Routledge.

Sorenson, David. 1998. *Shutting Down the Cold War: The Politics of Military Base Closure.* New York: St. Martin's Press.

Stedman-Edwards, Pamela. 1998. *Socio-Economic Root Causes of Biodiversity Loss: An Analytical Approach.* Washington, D.C.: Worldwide Fund for Nature.

Steigman, D. 1996. "Is It 'Urban' or Is It 'Asthma?'" *Lancet* 348: 143–144.

Stewart, Taimoon, and Dennis Pantin. 1991. "A Detailed Scrutiny in the Context of the NGOs' Vision of a Better Caribbean." Document presented at the Regional Economics Conference, Trinidad, February.

Stolberg, Sheryl Gay. 1999. "Poor Fight Baffling Surge in Asthma." *New York Times*, October 18.

Stonich, Susan. 1998. "The Political Ecology of Tourism." *Annals of Tourism Research* 25, no. 1: 25–54.

Switzer, J. V. 2001. *Environmental Politics.* Third ed. Boston: Bedford/St. Martin's.

Taylor, Frank Fonda. 1993. *To Hell with Paradise.* Pittsburgh: University of Pittsburgh Press.

Terrero, Ariel. 1996. "El contrato de la tierra." *Bohemia* (Havana) 88, no. 11: B8–B11.

Thrupp, Lori Ann, and Inelfe Pérez. 1989. "Breaking Chemical Dependency in Agriculture: The Remarkable Case of Cuba." Unpublished manuscript, Energy and Resources Group, University of California, Berkeley.

Toledo, Victor. 1992. "Utopia y Naturaleza: El nuevo movimiento ecológico de los campesinos e indígenas de América Latina." *Nueva Sociedad* (Caracas) 122.

Torrecilla, Arturo. 1986. "Por los caminos de la utopía: Ecología, vida cotidiana y nuevos protagonistas sociales." *Revista de Ciencias Sociales* 25, no. 1–2 (January-June).

Torres-González, A., and R. M. Wolansky. 1984. "Planning Report for the Comprehensive Appraisal of the Ground-Water Resources of the North Coast Limestone Area of Puerto Rico." U.S. Geological Survey Open-File Report 84–427.

Torres Vila, Cary, and Niurka Pérez Rojas. 1995. "La apertura de los mercados agropecuarios: impacto y valoraciones." Paper presented at the XIX Congress of the Latin American Studies Association, Washington, D.C., September 28–30.

Trinidad, Pablo. 2000. "Por una vía más moderada la economía." *El Nuevo Día*, April 16, Negocios, 9–11.

Trueba González, Gerardo. 1995. "Potencialidades del desarrollo agroindustrial cubano."

Paper presented at the XIX Congress of the Latin American Studies Association, Washington, D.C., September 28–30.

Tugwell, Rexford G. 1977. *The Stricken Land: The Story of Puerto Rico.* New York: Greenwood Press.

Turits, Richard. 2003. *Foundations of Despotism: Peasants, the Trujillo Regime, and Modernity in Dominican History.* Stanford, Calif.: Stanford University Press.

UMET (Universidad Metropolitana), New Jersey Institute of Technology, and the Centro de Acción Ambiental. 2000. *Resumen de Estudios y Datos Ambientales en Vieques,* San Juan, P.R.

United Church of Christ Commission for Racial Justice. 1987. *Toxic Wastes and Race in the United States: A National Report on the Racial and Socio-economic Characteristics of Communities with Hazardous Waste Sites.* New York: Center for Policy Alternatives, National Association for the Advancement of Colored Peoples and the United Church of Christ Commission for Racial Justice. www.ucc.org/justice/environment.htm.

United Nations Development Program (UNDP). 2003. *Human Development Report: Millennium Development Goals: A Compact among Nations to End Human Poverty.* New York: Oxford University Press.

United Nations Environmental Program (UNEP). 2000. GEO—Global Environmental Outlook. www.unep.org/geo2000/english/0183.htm.

Universidad Metropolitana, Instituto de Educación Ambiental. 2001. *Puerto Rico en Ruta hacia el Desarrollo Inteligente.* Executive summary. San Juan, Puerto Rico.

Urry, John. 2002. *The Tourist Gaze.* 2d edition. London and Thousand Oaks, Calif.: Sage Publications.

U.S. Census Bureau. 2000. *Census of Population 2000—Puerto Rico* and Junta de Planificación, Programa de Planificación Económica y Social, Oficina del Censo, San Juan, P.R.

U.S. Department of Agriculture, National Resources Conservation Service. 2000. *National Resources Inventory.*

U.S. Department of Health, Education and Welfare. 1954). *Vital Statistics of the United States, 1950.* Washington, D.C.: U.S. Department of Health, Education, and Welfare.

U.S. Environmental Protection Agency, Office of Policy, Planning, and Evaluation. 1992. *Environmental Equity: Reducing Risk for All Communities.* EPA230-R-92-008. Washington, D.C.: GPO.

U.S. Environmental Protection Agency. 2003. EnviroMapper website. http://map5.epa.gov/scripts/.escrimap?nam…MapperNbf&Cmd=ZoomInByZip&fipsCode=11220. Accessed August 2003.

U.S. House of Representatives. 1981. Committee on Armed Services. *Report of the Panel to Review the Status of Navy Training Activities on the Island of Vieques, Puerto Rico.* 96th Congress, 2nd session, no. 31.

U.S. Navy. 1999. *The National Security Need for Vieques.* Archived at www. center forsecuritypolicy.org; see also www.viequeslibre.addr.com/articles/navy_profit.htm.

Valdés Pizzini, Manuel. 1990. "Fishermen Associations in Puerto Rico: Praxis and Discourse in the Politics of Fishing." *Human Organization* 49, no. 2: 164–173.

Valdés Pizzini, Manuel, Ruperto Chaparro, and Jaime Gutiérrez. 1991. *In Support of Marine Recreational Fishing: An Assessment of Access and Infrastructure in Puerto Rico and the U.S. Virgin Islands.* Mayagüez: University of Puerto Rico Sea Grant College Program Research Report, PRU-T-91-001.

Valdés Pizzini, Manuel, Jaime Gutiérrez Sánchez, and Martín González. Forthcoming. Working Paper from the Centro de Investigación Social Aplicada (CISA), Departamento de Ciencias Sociales, Recinto Universitario de Mayagüez, and the University of Puerto Rico Sea Grant College Program.

Valdés Pizzini, Manuel, Juan M. Posada, Iván López, and Damaris Cabán. 1996. "Cognitive Constructions of Fishery Resources among the Fishers of Southwestern Puerto Rico." In *Proceedings of the Forty-Ninth Annual Gulf and Caribbean Fisheries Institute*, ed. R. LeRoy Creswell. www.gcfi.org, or contact Florida Fish & Wildlife Conservation Commission, Marine Research Institute, Marathon, Fla.

Velez-Arocho, Javier. 1994. "La calidad de nuestras aguas." *Boletín Marino* (University of Puerto Rico Sea Grant College Program), March-April.

Villamil, Joaquín. 1994. "Impacto de la construcción en la economía de PR." *El Nuevo Día* (San Juan, Puerto Rico), Suplemento Especial, September 3, p. S14.

Viola, Eduardo. 1992. "El ambientalismo brasileño. De la denuncia y concientización a la institucionalización y el desarrollo sustentable." *Nueva Sociedad* (Caracas) 122.

Vogel, Gretchen. 1997. "Why the Rise in Asthma Cases?" *Science* 276 (June): 1645.

Walling, Douglas, Mason, and Chevannes-Creary. 2004. *Caribbean Environmental Outlook*. Special Edition for the Mauritius International Meeting for the 10-Year Review of the Barbados Programme of Action for the Sustainable Development of Small Island Developing States. In UNEP, CARICOM, ed. Sherry Heileman. www.unep.org/geo/pdfs/Caribbean_EO.pdf.

Webb, Richard M. T., and Fernando Gómez Gómez. 1998. "Trends in Bottom-Sediment Quality and Water Quality in the San Juan Bay Estuary System, Puerto Rico." National Water Quality Monitoring Council, *Proceedings of the 1998 Conference*, Reno, Nevada.

Weisgall, Jonathan. 1994. *Operation Crossroads: The Atomic Tests at Bikini Atoll*. Annapolis, Md.: Naval Institute Press.

Williams, Raymond. 1973. *The Country and the City*. New York; Oxford University Press.

Winnick, Louis. 1990. *New People in Old Neighborhoods: The Role of New Immigrants in Rejuvenating New York's Communities*. New York: Russell Sage Foundation.

World Bank. 1993. *Caribbean Region: Current Economic Situation, Regional Issues, and Capital Flows, 1992. Country Study*.Washington, D.C.: World Bank.

———. 1997. *World Development Report: The State in a Changing World*. New York: Oxford University Press.

World Commission on Environment and Development (WCED). 1987. *Our Common Future*. New York: Oxford University Press.

World Resources Institute. 1996. *World Resources: A Guide to the Global Environment— 1996–97*. New York: Oxford University Press.

Wright, Angus. 1990. *The Death of Ramón González: The Modern Agricultural Dilemma*. Austin: University of Texas Press.

Yang, Andrea E. 2000. *Waste and Space–Beyond the NIMBY Discourse: The Political Ecology of Solid Waste in Santiago de los Caballeros, República Dominicana*. Master's thesis, Department of City and Regional Planning. Cornell University.

Yen, Irene, and S. Leonard Syme. 1999. "The Social Environment and Health: A Discussion of the Epidemiological Literature." *American Review of Public Health* 20: 287–308.

Zimbalist, Andrew, and Claes Brundenius. 1989. *The Cuban Economy: Measurement and Analysis of Socialist Performance*. Baltimore: The Johns Hopkins University Press.

NOTES ON CONTRIBUTORS

SHERRIE BAVER teaches political science and Latin American studies at City College and the Graduate Center-CUNY. Her work focuses on the Hispanic Caribbean and their mainland diasporas. Along with many articles, she has authored *The Political Economy of Colonialism: The State and Industrialization in Puerto Rico* (1993) and coedited *Latinos in New York: Communities in Transition* (1996).

MAURICE BURAC teaches geography at the Universite des Antilles et la Guyane in Martinique where he also heads GEODE, the Center for Research on Geography, Development, and Environment in the Caribbean. He has written extensively on Caribbean topics, and his most recent publication is *Les Antilles et la Guyane française à l'ube de XXIé siècle* (2003).

BARBARA DEUTSCH LYNCH is director of the urban and regional studies program and visiting associate professor in the department of city and regional planning at Cornell University. She has published on the politics of irrigation in Latin America, urban environment and the politics of risk, megaprojects and displacement, and on Latino environmental perspectives. Her recent research has focused on fields of urban environmental concern, their relationship to broader landscape transformations, and their implications for the distribution of risk in Caribbean cities. She came to Cornell from the Ford Foundation where she was program officer for the Caribbean environment and development program. She also taught in the Carleton College science and technology studies program and served as extension associate in Cornell's Institute for Comparative and Environmental Toxicology.

NEFTALI GARCÍA-MARTÍNEZ PH.D., a chemist, serves as executive director and chairman of the board of Scientific and Technical Services, a non-profit organization, where he also works as a scientific and environmental consultant. For over thirty-five years he has worked as an environmental advisor in Puerto Rico. His specialties include the areas of renewable and non-renewable natural resources, water, soil and air pollution, dangerous chemical substances, solid wastes and recycling, energy, reforestation, and matters of occupational health and industrial pollution. During the years as an advisor and community organizer, Dr. García has written numerous articles on the environment, many of which have been published in local newspapers, and is the author of *¿Quién cantará por las aves?* *("Who will sing for the birds?")* (1996), a collection of essays on the Puerto Rican environment. He is also a university professor and has taught chemistry, biochemistry, environmental conflicts, economic history of Puerto Rico, and environmental geography at several universities in Puerto Rico and the United States.

TANIA GARCÍA-RAMOS received her Ph.D. in social psychology from Universidad Complutense in Madrid and is now associate professor in the department of psychology, University of Puerto Rico. She has served on the board of directors of Scientific and Technical Services, an environmental NGO in Puerto Rico, since 1997, and since 1996, as a collaborator of the Taller Salud, a women's health NGO in Puerto Rico. She has edited two books and has authored several articles in her main areas of interest, women's health and environmental health.

FRANCINE JÁCOME serves as director of Instituto Venezolano de Estudios Sociales y Políticos (INVESP) in Caracas and teaches at Universidad Central de Venezuela. She is a well-known authority on Venezuelan politics, Latin American regional integration, and environmental movements and has written and lectured widely on these topics.

KATHERINE T. McCAFFREY teaches anthropology at Montclair State University, New Jersey. Her book, *Military Power and Popular Protest: The U.S. Navy in Vieques, Puerto Rico* (2002) analyzes the long-term conflict between the U.S. Navy and residents of Vieques Island. She continues to conduct ethnographic research on social change and the struggle for sustainable development on Vieques as the island shifts from military to civilian control.

MARIAN A. L. MILLER passed away in November 2003, and she was associate professor of political science at the University of Akron. Her work was at the intersection of development studies and global environmental studies, and she published numerous book chapters and articles. Her 1996 book, *The Third World in Global Environmental Politics* (Lynne Rienner), won the International Studies Associations' Sprout Award. Marian Miller will also be remembered as one of the founding associate editors of the journal, *Global Environmental Politics*.

LORRAINE C. MINNITE teaches American and urban politics at Barnard College, Columbia University. Her research focuses on issues of inequality, political participation, social movements, and institutional change. She has published work on a range of topics, including ethnic politics, social capital, voting, and immigration; and conducted several large-scale surveys and exit polls, working with unions and local community organizations. Currently, she is engaged in research for a book on immigrant rights in the United States.

IMMANUEL NESS is professor of political science at Brooklyn College-CUNY and associate director of the University's Graduate Center for Worker Education. Since 1999 Ness has edited *Working USA: The Journal of Labor and Society*. He has authored or coauthored numerous articles and books, most recently, *Trade Unions and the Betrayal of the Unemployed* (1998), *Central Labor Councils and the Revival of the American Unionism* (2001), and *Immigrants, Unions and the New U.S. Labor Market* (2005).

ANA RIVERA is completing a master's degree in environmental planning from University of Puerto Rico. At present, she works as an environmental planner at Scientific and Technical Services, Inc. Her areas of interest are the social and economic aspects of pollution and environmental degradation.

RICARDO "RICK" SOTO-LOPEZ has been in the planning profession for more than twenty years, and has extensive experience in redevelopment planning, community, and economic development. He has worked with a wide range of planning and community development organizations in New York, New Jersey, California, and more recently central Florida. As the senior planner for housing and community development with the city of Winter Park, Florida, he has facilitated the development of Hannibal Square Community Land Trust to foster long-term affordable housing in that city.

MANUEL VALDÉS PIZZINI is professor of anthropology and sociology, is the associate dean for research at the College of Arts and Sciences of the University of Puerto Rico in Mayagüez. He has also been director of the Puerto Rico Sea Grant Program, director of the Center for Applied Social Research, and research coordinator for the Caribbean Natural Resources Institute (CANARI). His research interests are human processes in the coast, uses of the forests and protected areas, and the historical dimension of resource utilization. Most of his research focuses on fisheries and fisheries management issues. He is the coauthor, with David C. Griffith, of the book *Fishers at Work, Workers at Sea: A Puerto Rican Journey through Labor and Refuge* (2002).

INDEX

academic community, and Puerto Rican
environmental movement, 55
access issues, 165; beach access, 53–54,
68; health care services, 149–150;
health insurance, 148–149, 149 *table*;
and land, 13; of resources and shelter,
168
acid rain, in Puerto Rico, 78
Acre Rubber Tappers' Movement, 111
adolescents, Puerto Rican, 83
Agauda, 44
"Agenda 21," 4, 19
agribusiness: and environmental decision
making, 169; threat of, 158
agriculture: Caribbean, 93–95; decline
in, 65; demise of, 54; export, 86, 106;
global nature of, 165; hillside, 107n.
1; in Martinique, 65–66; modern, 87;
in municipality of Rio Grande, 53;
plantation, 88, 159; in Puerto Rico,
48, 77–78; sustainable, 8, 86; UBPCs
in, 100; urban, 100–101, 106
agriculture, alternative, 88–89; in Cuba,
97, 98, 106–107; cultural obstacles to,
96–101; in Dominican Republic, 101–
104, 106–107; and global economy,
93–95; and Green Revolution
paradigm, 89–90; and rural land
tenure, 90–91
agriculture, sustainable: achieving, 105;
city on hill strategy of, 102–103; in
Cuba vs. Dominican Republic, 104–
105; obstacles to, 104; participatory

research approaches in, 103–104;
technology in, 99; transition from
conventional to, 105, 106; and withdrawal of land from agriculture, 91–93
agritourism, 106
agro-ecological experiments, 102–103
agro-ecosystems, 87
agro-forestry systems, 107n. 1
Aguirre Sugar Co., 127n. 5
air pollution: in Puerto Rico, 83–84. *See
also* pollution
allergies, 151. *See also* health problems
American Metal Climax, 79
Amis du Parc Naturel Régional, 69
ammunition depot, on Vieques, 109
animal traction, in Cuba, 99, 105
Anthony, Marc, 121
anti-Haitianism, 96
Antilles, 12
anti-tourism front, in Martinique, 67
Arbona, Sonia I., 51, 58, 59
architecture: colonial, 163; Jamaican-
Georgian, 42
arrabales, in Puerto Rico, 50
art, commodification of, 42
*Asociación de Desarrolla de San José de
Ocoa,* 101
Aspin, Sec. Les, 116
*Association de Défense du Patrimoine
Martiniquais et des Mal Loges*
(APPELS), 69, 70, 71, 72
*Association des Professeurs de Biologie
ou de Géologie* (APBG), 71